Lincoln's Generals

Lincoln's Generals

Edited by
GABOR S. BORITT

Essays by

STEPHEN W. SEARS
MARK E. NEELY, JR.
GABOR S. BORITT
MICHAEL FELLMAN
JOHN Y. SIMON

New York *Oxford* Oxford University Press 1994

Oxford University Press

Oxford New York Toronto
Delhi Bombay Calcutta Madras Karachi
Kuala Lumpur Singapore Hong Kong Tokyo
Nairobi Dar es Salaam Cape Town
Melbourne Auckland

and associated companies in
Berlin Ibadan

Published by Oxford University Press, Inc.,
200 Madison Avenue, New York, New York 10016

Oxford is a registered trademark of Oxford University Press

Library of Congress Cataloging-in-Publication Data
Lincoln's generals / edited by Gabor S. Boritt ;
essays by Stephen W. Sears . . . [et al.].
p. cm. Includes bibliographical references (p.)
ISBN 0-19-508505-1
1. United States—History—Civil War, 1861–1865—Campaigns.
2. Lincoln, Abraham, 1809–1865—Military leadership.
3. United States. Army—History—Civil War, 1861–1865.
4. Generals—United States—History—19th century.
I. Boritt, G. S., 1940– . II. Sears, Stephen W.
E470.L72 1995 973.7'3—dc20
94-11

9 8 7 6 5 4 3 2 1

Printed in the United States of America
on acid-free paper

1994
FOR DICK AND LEW
RENAISSANCE MEN
N. BORITT

By June it is usually warm, even hot, in Gettysburg. In the night at our farm, the fireflies glow in the dark, fleeting specks illuminating the woods and turning Marsh Creek into a pageant. In the daytime along the side of the road, orange tiger lilies proclaim their eternal message. My heart overflows; it is time to see old friends again, time to make new ones; it is the time for the Gettysburg Civil War Institute.

Farm by the Ford
Gettysburg

Gabor Boritt

Contents

Illustrations

Introduction

Gabor S. Boritt

THE MARAUDING YANKEES destroyed the Confederate pontoon bridge over the Potomac. No matter. July is a drought month when mighty rivers meander ever more sluggishly toward August. Then rivers are forded easily. In 1863, however, it was different. By the second week of mid-summer's month, the heavy rains swelled the Potomac high above its usual course. "Had the river not unexpectedly risen," Robert E. Lee wrote to his wife Mary Custis, "all would have been well with us. But God, in His all-wise providence, willed otherwise . . . ."[1] Having lost a third of its men at Gettysburg, the Army of Northern Virginia faced the possibility of disaster.

In Washington, Abraham Lincoln knew this. He thought of going to the Potomac by Falling Waters and Williamsport to take command of the Union forces. But he resisted temptation. His mind's eye pictured a raging, cresting river that spelled the doom of Lee's soldiers and of the Confederacy. The Army of the Potomac held the Army of Northern Virginia, and so the war, "in the hollow of their hands."[2]

The commander-in-chief spent ten agonizing days using all his spirit, eloquence, and guile trying to force his general, George G. Meade, to close this army's hand. Lincoln failed. Eleven days after Gettysburg, the president felt utterly defeated—defeated by his own generals. He was left in tears. Lee had escaped.

General Meade, of course, saw the Gettysburg campaign very differently. This modest and brave man would not say, but his admirers said it for him, boldly: he had saved the American Republic. In 1865 Harvard University would bestow an honorary Doctor of Laws degree on him with the following citation:* "Illium exercitus Americani imperatorem, qui periculosissimo belli discrimine res patriae virtute et consilio restituit, Georgium Gordon Meade." Even in the July heat of 1863, sharply diverging views notwithstanding, general and commander-in-chief reassured each other with expressions of high regard. But strain formed a central part of their relationship. And to a greater or lesser degree, the same could be said about Lincoln and his other generals.

*"We honor him, the commander of the American army, who in the most perilous crisis of the war, restored the fortunes of his fatherland by his courage and counsel: George Gordon Meade." I am indebted to Professor Charles J. Zabrowski of Gettysburg College for helping with the translation from the Latin.[3]

Many elements mixed to augment the partly inherent conflict between civilian and military authority. Warfare evolved, and tactics, strategy, weapons, who should do what and how, were all up for grabs. Politics and personalities entered the picture. So did differences in constituencies, and hence responsibilities; often divergent war goals; and ultimately divergent conceptions of what the United States was and should be.

This book takes a hard look at the interaction of five leading generals with their Civil War commander-in-chief. The choice of the five lieutenants no doubt colors its findings. Surely, future students will look at others, too, as well as the five here. Different eyes see different things, and that is all to the good.

In Chapter One, Stephen Sears, indisputably the leading expert on George Brinton McClellan, brings together the fruit of long years of learning. Sears explains that with no other commander did Lincoln get so closely involved. At a time when the president had to go through a forced education in military science, McClellan served—for good and for ill—as his most influential teacher. This soldier grew to be *the* looming presence as Lincoln managed the war during its first year and a half or so. At the age of thirty-four, "little Mac" took command of the Union's largest army and continued in command for more than a year. He also served as general-in-chief through the first winter of the war.

From before the Peninsula campaign to beyond Antietam, Lincoln expended untold effort to make their partnership work. He asserted himself as the strong commander-in-chief and also utterly humbled himself before his haughty soldier. He did all and endured all. But he failed. "Lincoln could do everything for this general but make him fight," Sears writes, "and in the end, that is the measure of the general. George McClellan was simply in the wrong profession. He might have been highly successful in many other fields . . . but field command in a civil war was not one of them." In fact, the fame garnered in a victory-starved North propelled McClellan into the 1864 presidential race. He then challenged the man who had fired him from command after Antietam. He lost again.

If Lincoln developed some sterling qualities as commander-in-chief, these did not always serve him well, Mark Neely suggests in Chapter Two. "Fighting Joe" Hooker has found few champions among latter-day historians, and Lincoln's relationship with him is sometimes portrayed in one-sided terms, in those of stern and wise father to son. In his famed "Hooker letter," the president told the general that he had been given command of the Army of the Potomac not because of but "in spite of" having said that the nation needed "a Dictator." "Only those generals who gain success can set up dictators. What I ask of you is military success, and I will risk the dictatorship."[4]

Neely by-passes such well-known material as he seeks out new insights in the context of the battle of Chancellorsville and the Lincoln-Hooker connection. Terrain, he argues is "perhaps the most underestimated factor" in twentieth-century military writing about the Civil War. The Wilderness was poorly suited to attack by the federal forces, "whether led by the inadequate Hooker or by the brilliant Ulysses S. Grant." When Lincoln nonetheless urged attack, he failed his general. The same can be said about the commander-in-chief's unwillingness to protect Hooker, or other chief lieutenants, from exposing themselves needlessly on the battlefield. Hooker's bow toward the nineteenth century's cult of manliness led to his wounding at Chancellorsville. He was "a perfect example of the unnecessarily brave Civil War general." Thus, in part, Neely looks at his subject in the light of gender studies. In part, he faults military historians for accepting the "myth of Braddock's defeat"—a belief in the superior American ability to fight in the woods, of warrior ways learned from the Native Americans. Hooker proved at Chancellorsville what myths, and the failure to adapt to wooded terrain, could bring: devastating defeat.

If Lincoln's emphasis on attack served him poorly in the Wilderness, Neely nonetheless sees this deeply rooted presidential characteristic as a virtue. For the commander-in-chief recognized "the imperative of hard fighting in an era before the germ theory of disease lessened the risks of long campaigns." Indeed, as the Gettysburg campaign moved towards a potential climax on the banks of the Potomac, Lincoln again desperately urged attack.

Had George Gordon Meade died on July 4, 1863, he would have become one of the greatest heroes of the Civil War. But he did not die on that long-ago Independence Day. And so Chapter Three looks at the Meade-Lincoln connection of history. As I wrote this chapter, I found myself driving daily on the Fairfield Road, from my farm to Gettysburg College. Seminary Ridge now has a traffic light; and as I would come close to it, I would hurry, hoping to make the light. Sometimes I found a deliberate driver ahead who would slow down, and then slow down some more, making absolutely sure that the crossroad was safe. We would both miss the light. And I would think of the driver in front of me as General Meade.

Perhaps such a glimpse into an author's mind helps readers, though it should be clear that driving is no analogy to war. Yet Meade, like McClellan, had "the slows." At Gettysburg he decisively defeated the Army of Northern Virginia, for the first time ever. Immediately after, he again faced Lee at his best, but equally important in what followed was the kind of person Meade was. Indeed, the Army of the Potomac would not fight another major battle until Ulysses S. Grant took over in the spring of 1864.

When Lee came north in the June of 1863, the North panicked, while Lincoln quietly exulted. He saw an opportunity to destroy the Confederacy. After Gettysburg, the North exulted; but Lincoln refused to use the word victory, and soon penned to Meade an unsent letter: "My dear general, I do not believe you appreciate the magnitude of the misfortune involved in Lee's escape."[5] Lincoln and Meade had gone at each other: the soldier cautious; the commander-in-chief aggressive and often misinformed about both big and small matters, but ultimately right in his thinking. The two could not reconcile their differences. They approached the war and the world in ways all too far apart. Though inconvertible proof is not possible, and the would-have-beens of history are suspect, Chapter Three suggests that Meade should have been more aggressive from July 4 on, pursued Lee faster, hazarded attack well before July 14 with its immense attendant casualties, and failing in all that should not have allowed the Confederates to streak across the river practically unmolested. But in one sense this is idle speculation. Given who Meade was—"a *juste milieu* man," a middle of the roader, as he described himself as early as 1846—he could do no other.[6]

Lincoln sensed this, at moments knew this, deepening his helpless frustration. In mid-July there had been a chance that military necessity might send him to Gettysburg; political necessity did send him in November. However deep his old hurt about the "unfinished work" of Gettysburg, there he spoke his few words that helped turn Meade's battle into American legend.

Lincoln's relationship with Meade had been, at moments, almost unbearably intense; with General William Tecumseh Sherman, it was remote and intermittent, but nonetheless revealing. The Lincoln-Sherman connection, Michael Fellman shows in Chapter Four, also shed important light on diverging views about the future of the United States.

Lincoln and Sherman met only three times: twice in 1861 and once in 1865. In between, they corresponded infrequently about minor matters, the president usually making requests for some "kindness to Southern civilians, which Sherman would reject." But the general also disregarded and even publicly opposed the president on bigger issues. Most important, Lincoln started moving toward the use of black troops in 1862 and went public with his goals in the Emancipation Proclamation. When the implementation of the policy reached Sherman's command, he balked. He reached the point of insubordination. "I like niggers well enough as niggers," he wrote to a friend, "but when fools and idiots try and make niggers better than ourselves, I have an opinion."[7] Accordingly, the general implemented his own views. And because he was a "winner," indispensable to the war effort, the commander-in-chief put up with the disloyalty. Indeed, Sherman's capture of Atlanta helped ensure Lincoln's re-election. The March to the Sea and beyond helped ensure final Union victory. Thus, ironically, Sherman's military prowess contributed significantly to the triumph of a policy he despised. Lincoln was unstinting in his praise: "The honor is all yours," he wrote Sherman after the fall of Savannah. The relationship of the president and general was thus "both antagonistic and symbiotic." Or to use the colloquial, it resembled at times the "good cop, bad cop" approach. "But this," Fellman writes, "was the unintended outcome of two distant and disparate men."

Lincoln's most able lieutenant was Sherman's mentor, Ulysses S. Grant. The man known as "Unconditional Surrender" Grant went east to take command of all the armies of the United States in the spring of 1864, and in the years after much would be made of the mutual respect, harmony, and trust that had grown between him and the commander-in-chief. Grant himself contributed substantially to the sentimentalization of this relationship. But John Simon, the most knowledgeable student of the general shows in Chapter Five that contemporary documents bespeak a different story.

Grant had no interest in politics; indeed, he had an aversion to it. All the same, Lincoln not only gave strong support to the Grant of the battlefields, but also kept a wary eye on him as a possible political player. Eighteen sixty-four was an election year; and a prewar Democrat, now war hero, was a potential threat. After the president attained renomination in mid-year, his watchfulness abated somewhat, especially when the lieutenant-general offered unconditional support for the re-election campaign.

Grant, in turn, also kept a vigilant eye on the White House. The fate of the first commander of the Army of the Potomac had become a cautionary tale among the generals, and Grant did not intend to be "McClellanized." He watched and kept his distance.

Even if Grant thought that the control of the war passed into his hands, Simon writes that Lincoln "gave his bulldog a longer leash than previous commanders, but a leash nonetheless." After the Army of the Potomac again bogged down at Richmond, and the Confederates simultaneously managed to reach the outskirts of Washington, in mid-1864, the commander-in-chief quickly showed who was in charge. "In effect," he told Grant "where to go," "what to do," and "at least hinted what might happen if he failed in the assignment."

Together thus, they moved toward victory at last. Tensions continued between them but, as with the earlier lieutenants, so did concert. Minor disgraceful episodes also intruded into their relationship. While still in the West, Grant expelled Jews from his command, and Lincoln revoked the order. Later, in the East, Lincoln ignobly asked that his son be appointed to Grant's staff; it was done and *Captain* Robert Lincoln was created. But the war went on.

When the end came into sight, Lincoln, not Grant, pushed for a policy of unconditional surrender, popular sobriquets notwithstanding. Unequivocal orders went to the general forbidding peace negotiations with the Confederates. And Grant learned well. Even after Lincoln's assassination, Lincoln's policy ruled. The lieutenant-general stood firmly against Sherman, insisting that only the Lincolnian policy of unconditional surrender was acceptable. The implications were enormous. For the moment, the future looked promising for the United States.

I am indebted to many people for helping to create *Lincoln's Generals*. My co-authors, one and all, proved to be a pleasure to work with. At the 1993 Gettysburg Civil War Institute session, Tina Fair, Linda Marshall, and Martie Shaw again did their best while the historians presented these essays. Some three hundred students, ages sixteen to the eighties—though a couple of twelve-year-old geniuses were also present—supported us with probing questions and good cheer. Much needed help also came from our fine student assistants: Jen Haase, Susan Fiedler, Steve Petrus, Cameron OBrion, Al Pennino, Tracy Schaal, Pat Taylor, Pete Vermilyea, and two of my sons, Jake and Dan Boritt. The unflappable Dr. William Hanna kept our sizable group of scholarship students on their best learning behavior.

I had asked each of the historians contributing to this volume to provide suggestions for further readings. Mark Neely responded with much more, a wonderful historiographical essay. It does provide bibliographical suggestions for the readers, but deserves inclusion in this book on its own right, too.

Elizabeth Lincoln Norseen Boritt, my wife and closest friend, helped always. Norse, our oldest son, designed the dedication page. Jake, our middle son, created his father's flattering portrait for the dust jacket. And like our youngest, Daniel, they shared their joy in life.

Martie Shaw, ever well-organized and cheerful, got the manuscript ready for publication.

My editors at Oxford, Sheldon Meyer and Leona Capeless, proved once again to be delightful co-workers.

This book originated in a conversation with Joseph Glatthaar in the early autumn of 1991 while he served as the Harold Keith Johnson Visiting Professor at Gettysburg's neighbor, the U.S. Army War College in Carlisle. I am grateful. When I mentioned a plan to focus an upcoming Civil War Institute session on Lincoln as a military leader, he countered that we should examine relations between commander-in-chief and principal generals. So it was to be.

Lincoln's Generals is dedicated to Richard Gilder and Lewis Lehrman, learned and generous friends, co-founders of the Lincoln Prize.

1

Lincoln and McClellan

Stephen W. Sears

LINCOLN AND MCCLELLAN

Just before this photo was taken in 1862, Lincoln wrote to his wife Mary: "Gen. McClellan and myself are to be photographed . . . if we can be still long enough. I feel Gen. M. should have no problem. . . ." *(Photograph by Alexander Gardner, courtesy of James Mellon. Text of the Lincoln letter courtesy of Lloyd Ostendorf)*

ON OCTOBER 29, 1985, A TUESDAY, a dozen or so self-described "Civil War buffs and historians" assembled on Connecticut Avenue at California Street in Washington. The site was the equestrian statue of Major-General George Brinton McClellan. The occasion was the 100th anniversary of the general's death. Several in the group wore uniforms of Federal blue. One wore Confederate gray, in acknowledgment of the fact that one of the pallbearers at General McClellan's funeral had been his old friend and foe, General Joe Johnston. The ceremony was brief but heartfelt and, as such things do in a city like Washington, it attracted a small crowd of curious spectators. A wreath of laurel tied with red, white, and blue ribbon was laid at the base of the statue, and there was a short address "touching on some of McClellan's strengths." "We concluded," goes the report, "with three cheers for 'Little Mac,' which echoed down the avenue and over the traffic. . . ."[1]

Washington statuary honors a number of Civil War generals, of course—among them Hancock, Thomas, Sheridan, McPherson—and several have their names on circles or squares as well, but McClellan's sculptor was uniquely successful in capturing what a splendid figure the general made on horseback. The setting is also noteworthy. General McClellan, appropriately enough, is posed looking resolutely southward, while to the rear are ample avenues by which he might change his base if need be.

The gallant band of McClellan loyalists gathered on Connecticut Avenue that day was but a faint echo of the reaction to the general's death one hundred years earlier, in 1885. Then the news was on front pages across the nation. President Grover Cleveland sent condolences. The press carried statements from titans of finance and captains of industry, from such ranking generals of the late war as Sherman, Hancock, and Sheridan, Beauregard and Joe Johnston. There were lengthy obituaries and editorial-page notices. Some of these were reflective; others were written with a partisanship still stirring emotions two decades after the close of the war. The *Philadelphia Ledger* expressed a widely held view of General McClellan. "Concerning his abilities and merits as a general commanding great military operations," the paper editorialized, "there has been hot contention, and there is likely to be wide divergence of opinion for at least a generation to come—possibly for all time."[2]

Invariably, any discussion of George McClellan, major-general, encompasses Abraham Lincoln, commander-in-chief. With no other wartime lieutenant was Lincoln so closely involved. During the first nineteen months of the war, McClellan was a vivid, virtually constant presence in the president's management of the contest. For fifteen of those months he commanded the North's largest army, serving also, during the first winter of the war, as general-in-chief of all the North's armies. Thereafter, for two years, McClellan was a principal player on the political stage.

The Pennsylvania editor Alexander McClure, in his *Abraham Lincoln and Men of War-Times,* remarked that "McClellan's ability as a military commander, and the correctness of Lincoln's action in calling him to command and in dismissing him from command, are as earnestly disputed to-day as they were in the white heat of the personal and political conflicts of the time."[3] McClure wrote that in 1892. Today, more than a century later, the earnest dispute is no longer at white heat, yet it may be worthwhile to stir the coals. The intent will be to generate light, not heat. Because the Lincoln-McClellan partnership was most intense during the evolution and execution of the Peninsula campaign, that campaign will merit our closest attention.

Once George McClellan formed his opinion about a subject or an individual, he seldom budged from it. He was prideful in his convictions, certain in his chosen course. The best evidence suggests that he formed his opinion of Abraham Lincoln on first acquaintance, in Illinois in the late 1850s, and that it was not favorable. At the time, McClellan was vice president of the Illinois Central, and Lincoln performed legal work for the railroad.

We know nothing of their personal relationship, but politically there was certainly no meeting of minds. McClellan was a conservative Democrat and an active supporter of Stephen A. Douglas in the 1858 Illinois senatorial contest, to the extent of giving Douglas use of his private railway car for campaigning. McClellan would describe Lincoln's performance in one of the debates with Douglas as "disjointed" and "rather a mass of anecdotes than of arguments." There was no comparing the two candidates in oratorical powers, he thought.

To be sure, this was McClellan's postwar recollection, when his views were embittered and set in stone. But there is as well contemporaneous evidence that he considered Lincoln a weak reed. In January 1861, a fellow Illinois Central director, John M. Douglas, described Lincoln for McClellan as "not a bold man. Has not nerve to differ with his party and its leaders. . . . O Mc you and I know L and we know that he can not face the opposition which would rise if he were to take the right stand. . . . I tell you it is impossible for him to lead. . . ." John Douglas wrote this in the confident assumption that George McClellan shared his view of the president-elect. The coming months would demonstrate the assumption to be correct.[4]

"Meteoric" is a word often used to describe McClellan's rise to high command after Fort Sumter, and after First Manassas. Eyebrows were raised at the Lincoln administration's promotion of a thirty-four-year-old erstwhile captain of cavalry to such high military postings with such indecent speed. At the time and place, however, there was eminent logic behind each promotion. It was Winfield Scott who put McClellan into play, and when we look over Scott's shoulder at the pool of officer prospects, it is easy to see why.

George Brinton McClellan was regarded as one of the best and brightest in the old army. Second in his West Point class of 1846, he entered the elite Corps of Engineers and served there in the elite Company of Engineer Soldiers. He did well in combat in the Mexican War and afterward commanded his company at West Point. He served General Scott in Washington and managed short-term engineering projects. He explored in Texas and for the Pacific Railroad Survey. He ran a secret mission to Santo Domingo to explore coup d'état prospects there. His plum army assignment was a mission to observe the war in the Crimea and to report on the armies of Europe. Ten assignments in nine years was hardly the norm in the antebellum army.

Resigning his commission in 1857, McClellan became one of the nation's highest paid railroad executives, first with the Illinois Central and then with the Ohio & Mississippi. These were not boardroom posts but involved hands-on experience in the operation of the roads, a useful sort of civilian employment for a prospective military executive. To be sure, McClellan's experience of military command was limited, but the same could be said of almost all newly minted general officers in 1861. Such was the regard for him in the army that when he married in 1860 and three years out of the army, General-in-Chief Winfield Scott attended the wedding.

In answering Lincoln's call to arms after Fort Sumter, the North's three most populous states, New York, Pennsylvania, and Ohio, competed for ex-Captain McClellan's services to command their state troops. He accepted Ohio's bid and on April 23 was made a major-general of volunteers. Just four days later, he sent General Scott his strategic plan "tending to bring the war to a speedy close." It was the first such plan by anyone, and offers a first glimpse at a major strain in George McClellan's military character. He never thought in lesser terms than ending the contest with a single massive thrust—under his singular leadership.

After securing the Ohio River line, McClellan proposed to assemble an 80,000-man field army. If only cooperation with the eastern army assembling at Washington was wanted, he would march "with the utmost promptness" through western Virginia to fall on Richmond. McClellan's real interest was reserved for a second and far more ambitious plan. To crush out secession in the Deep South, he would lead his 80,000 men due south on Nashville. Gaining a battle there and securing Kentucky and Tennessee, he would move against Montgomery, then the seat of the Confederate government. The eastern army, having meanwhile taken Charleston and Augusta, would then combine with his and end the rebellion by seizing Pensacola, Mobile, and New Orleans. "The 2nd line of operations could be the most decisive," said McClellan.

General Scott pointed out the shortcomings in the plan, particularly its failure to utilize the western theater's river network. The general-in-chief then responded with a strategic plan of his own, based on a naval blockade of the Confederacy's coasts and an advance (by McClellan's 80,000) down the Mississippi to New Orleans. It would be dubbed the "Anaconda Plan" for its emphasis on crushing the Confederacy in the coils of economic pressure. Despite the differences in the two plans, Scott applauded McClellan for his "intelligence, zeal, science, and energy" and named him major-general in the regular army. Only Scott himself now ranked him. McClellan's new title was commander of the Department of the Ohio.[5]

In his handling of events in Kentucky and western Virginia in the spring and summer of 1861, McClellan would reveal a second element of his military character. Political issues, particularly the issue of slavery, carried equal weight with military matters in his plans for ending the rebellion. "All your rights shall be religiously respected," he promised in a proclamation to the people of western Virginia. He pledged no interference with slavery; indeed, Union troops would put down "with an iron hand" any attempt at slave insurrections. To one of his lieutenants he spelled out the rules: "All private property whether of secessionists or others must be strictly respected, and no one is to be molested merely because of political opinions."

After issuing his proclamation, McClellan on May 30 wrote for the first time to President Lincoln. He had acted entirely on his own, he said, but not without "knowledge of your Excellency's previous course & opinions. . . ." He hoped he had read the president correctly, he said, for in going public with his proclamation, "I have not intimated that it was prepared without authority." If Mr. Lincoln had any thoughts about this presumption by one of his newest lieutenants, he kept them to himself.[6]

At that time and place, Lincoln had little reason to object to the general's handling of the slavery issue, except perhaps on the ground that the policy was being set by a general in the field who pledged the army to its enforcement. Meanwhile, McClellan's campaign in western Virginia was producing what few military successes the North could claim in these months. In this first encounter with General McClellan, Lincoln chose to watch silently from the sideline.

McClellan's advancement from head of the Department of the Ohio to head of the Department of the Potomac at Washington, following the First Manassas debacle, can be criticized only with the benefit of hindsight. McClellan outranked, and had clearly outshone, McDowell and his lieutenants in the eastern army. Frémont's commission as major-general in the regular army might bear the same date as McClellan's, but Frémont had yet to gain reputation or notoriety. No one in Washington seriously considered replacing McDowell with such major-generals of volunteers as Dix or Banks or Ben Butler. McClellan was the natural choice. He seemed the perfect choice.

The day after arriving in Washington to take command, McClellan wrote his wife, in a much-quoted letter, "I find myself in a new & strange position here—Presdt, Cabinet, Genl Scott & all deferring to me—by some strange operation of magic I seem to have become *the* power of the land. I almost think that were I to win some small success now I could become Dictator or anything else that might please me. . . ." He was not of course using dictator in today's sense of a tyrant, but rather in its older, benign Roman sense of rescuing the government in time of peril, or in the sense of a general like Belisarius responding selflessly to the people's appeal. Indeed, a wartime cartoonist would label a drawing of General McClellan "The Modern Belisarius."

Of more interest in this letter is McClellan's characterization of Lincoln as "deferring" to him. Lincoln's biographer Benjamin P. Thomas believed the president was highly effective in dealing with the people he encountered in his official capacity. "Meeting all sorts of people, he shaped his response to their approach," Thomas wrote, and he added, "Men of the strongest personalities felt Lincoln's quiet dominance." General McClellan proved an exception to this generalization.[7]

To be sure, much of what we know of Lincoln's dealings with McClellan comes from McClellan's testimony. In such matters, especially when the dealings were private, he is not always the best witness. Occasionally Lincoln might report on meetings with the general to his friend Orville H. Browning or to Gideon Welles, who recorded the remarks in their diaries. More often, however, the record of a Lincoln-McClellan private encounter is likely to be in McClellan's hand, in his letters to his wife or sometimes in letters to his New York confidant Samuel L. M. Barlow.

General McClellan reached Washington July 26, 1861, and he departed for the Peninsula on April 1, 1862, and during the intervening months the record shows Lincoln was with the general on at least fifty-seven occasions. The accounts by third parties of three of these occasions suggest something of their general character.

William Howard Russell, the *Times* of London's war correspondent, has left us a picture of one of Lincoln's frequent "drop in" visits to the general. One evening early in October 1861, Russell was in the anteroom at army headquarters on Jackson Square when in walked the president, "a tall man with a navvy's cap, and an ill-made shooting suit, from the pockets of which protruded paper and bundles. 'Well,' said he to Brigadier Van Vliet, who rose to receive him, 'is George in?' " The general was resting after a fatiguing day, Van Vliet explained, but would be told the president was there to see him. " 'Oh, no; I can wait. I think I'll take supper with him,' " Lincoln said and sat down to wait with the rest. "This poor President!" Russell remarked. ". . . He runs from one house to another, armed with plans, papers, reports, recommendations, sometimes good-humoured, never angry, occasionally dejected, and always a little fussy."

On another evening that fall, General Samuel Heintzelman encountered Lincoln and McClellan together in the headquarters map room. "The President continued to pore over a map," Heintzelman recorded in his diary, "—our map of the war on this side of the river, making remarks, not remarkably profound, but McClellan listened as if much edified. At last the President rose & said he was ready & they adjourned to an inner room, where the President spoke so loud we could not help hearing. They were engaged making their arrangements for Gen. Halleck to take command in the West." Afterward, Heintzelman wrote, McClellan saw the president out, "& as he pushed the door to looking back said 'Isn't he a rare bird.' "

The third of these incidents is the best known. Two days after the encounter described by Heintzelman, President Lincoln, Secretary of State Seward, and presidential secretary John Hay called unannounced on General McClellan in the evening at his house. The general was attending an officer's wedding, they were told; would they care to wait in the parlor? After an hour McClellan returned and, ignoring his orderly's announcement of the visitors, went upstairs to his quarters. After an interval the orderly was sent upstairs to remind McClellan of his visitors. He returned to say that the general had gone to bed. Hay entered in his diary, "I merely record this unparalleled insolence of epaulettes without comment." Lincoln was apparently unruffled, remarking only that this was not the time "to be making points of etiquette & personal dignity."[8]

Lincoln's habit of dropping by unannounced to take supper or to talk over events greatly annoyed McClellan, and his snub was calculated to break the habit. On another occasion, he confided to his wife that he was concealing himself at Edwin M. Stanton's—Stanton at the time was his friend and adviser—so as "to dodge all enemies in shape of 'browsing' Presdt etc." Yet there were times when McClellan seemed disarmed by the president's folksiness. "I have just been interrupted here by the Presdt and Secty Seward who had nothing very particular to say, except some stories to tell, which were as usual very pertinent & some pretty good," he told Mrs. McClellan on October 16. "I never in my life met anyone so full of anecdote as our friend Abraham—he is never at a loss for a story apropos of any known subject or incident." On more public occasions, however, McClellan found this presidential story-telling habit undignified. Of a Washington reception at which he met Lincoln he wrote, "was of course much edified by his anecdotes—ever apropos, & ever unworthy of one holding his high position."[9]

In describing the Lincoln of this period in an early draft of his memoirs, McClellan reflected an impression apparently first formed in his Illinois Central days. "He was not a man of very strong character," he wrote of the president, "& as he was destitute of refinement—certainly in no sense a gentleman—he was easily wrought upon by the coarse associates whose style of conversation agreed so well with his own."

The gentlemanly virtues were important to George McClellan. At the height of some dispute later in the war with General Halleck, he burst out to his wife, "He is not a refined person at all. . . ." This attitude obviously formed one of the barriers to Lincoln's efforts to reach General McClellan and to reason with him. "Lincoln wore no outward signs of greatness," Benjamin Thomas wrote. "He inspired no awe or embarrassment. He had no pomp, no wish to impress." In dealing with the patrician McClellan in so unaffected a manner, perhaps Mr. Lincoln miscalculated. Rather than deference, perhaps a more imperial attitude might have produced better results.[10]

However that may be, Lincoln continued to defer to the new commander of his principal army, christened by McClellan the Army of the Potomac. Winfield Scott remained in overall military charge of the war effort, continuing his punctilious relationship with the president. McClellan grew restive under Scott, then impatient, finally rebellious. Like much else about George McClellan, there was a pattern to this. Throughout his military and civilian careers he had become embroiled in disputes with his superiors or with anyone else he believed to be standing in his way.

McClellan's dispute with General Scott centered on their differing views of the enemy they faced. This perception of the enemy marks the third element of McClellan's military character, after his predilection for grandiose schemes and his delving into political issues. The origin of this third element can be pinpointed precisely—August 8, 1861. On that date he announced in a dispatch to Scott that 100,000 Confederates were poised for an attack on the Army of the Potomac, which was "entirely insufficient for the emergency."

McClellan credited "information from various sources" for this critically important first reading of the enemy he faced. The one source we know about is a Confederate deserter's statement of "the strength, plans &c of the Rebels." We know, too, that Allan Pinkerton, McClellan's intelligence chief, contributed nothing to this finding, for on this date he had not yet set up his intelligence operation. The estimate of a 100,000-strong Rebel army was entirely General McClellan's invention, and General Scott would have none of it.

And invention it was, a tripling of the actual size of the force he faced. From that day forward, 100,000 as the enemy's troop strength was a figure he would never—could never—retreat from. It could only grow larger, and did, to as many as 170,000 by mid-September. By the next spring, when he made his march on Richmond, the Rebel host had grown in his mind's eye to 200,000. And, of course, from the first day of his command to the last, it was his own army that was the larger one, by a substantial margin. By desperate effort, including unsupported speculations he labeled "general estimates," detective Pinkerton sought to keep pace with his chief. Often enough, McClellan reported to Washington other, even larger figures than those supplied by Pinkerton.

There can be no doubt that General McClellan accepted the inflated numbers without hesitation and without reservation, and that he based all his military decisions, both strategic and tactical, on them. His wholehearted acceptance of the estimates is confirmed by the number of times he mentions them in his letters to his wife. When writing to Mrs. McClellan, without exception he told her truths about everything he did and felt and believed, or at least the truths as he understood them. "In talking or writing to you," he once told her, "it is exactly as if I were communing with myself—you *are* my *alter ego*. . . ."

If it is clear enough that General McClellan believed the fiction that the Confederate army facing him in every campaign, in every battle, was considerably larger than his own, the question must be asked: Why did he believe this? The answer seems to be: Because of the logic of the case.

McClellan was certain that the South had been preparing for war and mobilizing its forces long before Fort Sumter. When he came to Washington to take command, the generals he faced—Beauregard, Joe Johnston, later G. W. Smith—were old and respected army friends who, along with McClellan's onetime patron Jefferson Davis, could be expected to assemble the great mass of their forces for an all-out effort to seize the Federal capital. Logic told McClellan this was what the Confederate leaders should do and would do; it was what he would do had roles been reversed. "Were I in Beauregard's place," he wrote in his August 8 warning, "with that force at my disposal, I would attack. . . ." And once the logic of the premise was firmly established, there was no way a man of McClellan's proud certitude could back away from it.[11]

No one stepped forward to challenge his figures or his reasoning, which served to further strengthen the myth.

General Scott said only, "I am confident in the opposite opinion. . . . I have not the slightest apprehension for the safety of the Government here." All other views of the enemy were by individuals and were unofficial and so carried little weight. The visiting Prince Napoleon of France inspected the Confederate army in its camps around Manassas and returned to Washington with an estimate of its numbers as 60,000 (which was considerably too high). Not long afterward, McClellan assured Secretary of War Cameron that the correct figure was 170,000.

It must be remembered that when the commanding general of the nation's largest army (and, after November 1, the commanding general of all the nation's armies) issued an official judgment of the enemy he faced, it was not a matter that invited debate. The army was the sole collector of intelligence; Allan Pinkerton considered himself the personal employee of General McClellan; and the general kept a tight rein on the whole process.

That Mr. Lincoln accepted McClellan's delusions at face value may be doubted. We know from the diaries of Welles and Chase that the matter was discussed in cabinet, at least at later times when the majority of the cabinet had grown disillusioned with the general, and questions were raised about numbers. In due course both Welles and Chase went on record as disbelieving McClellan's assertions of being forever outnumbered.

The president's views must be inferred, but clearly he grew wearied by McClellan's incessant calls for reinforcements. He remarked to Senator Browning that should he send McClellan the latest contingent of 100,000 reinforcements he called for, the general would say the next day that now the enemy had 400,000 and therefore he could not advance. To be sure, this was said after the Peninsula campaign, when the president's patience with his general had worn very thin. It seems certain that at least by 1862 Lincoln's highly developed strain of common sense led him to view McClellan's figures with a jaundiced eye. It is equally likely that the president, as Edwin C. Fishel believes, came to regard these exaggerations "as another McClellan effrontery that was easier to live with than to resolve." In any event, the administration had little with which to challenge these numbers had it wished to. In the intelligence game General McClellan held all the cards.[12]

In his first White House meeting with McClellan, on July 27, the president had asked the general for his thoughts on (as McClellan put it) "conducting the war on a large scale." A week later McClellan handed in his plan, and Lincoln had him read it to the assembled cabinet on August 3. It was essentially a textbook plan, delivered in broad strokes and derived from the general's extensive reading in the military literature. And it was grand beyond all previous plans. He would assemble and lead a field army of 273,000 men and 600 guns. An additional 100,000 would be required for garrison troops and a reserve. He set his objectives as "thoroughly defeating their armies, taking their strong places, and pursuing a rigidly protective policy as to private property"—a euphemism for slavery—"and unarmed persons. . . ."

McClellan made Richmond, now the Confederate capital, his first target, specifying no particular line of advance. Afterward the grand army would march southward on Charleston and Savannah, then sweep across the Deep

South, taking Pensacola, Mobile, Montgomery, New Orleans. The advance down the Mississippi, the centerpiece of General Scott's Anaconda Plan, was secondary in General McClellan's plan. It was a plan of Napoleonic scope, suited to a general the newspapers were calling "the Young Napoleon."

The life span of this grand plan was one week. On August 8, when McClellan raised the specter of an attack by the newly invented Confederate host at Manassas, the plan was put aside and not heard of again. Instead, McClellan devoted all his energies to defense, to strengthening the Army of the Potomac, to pushing General Scott—that "perfect imbecile," he called him—from his path. On November 1, Scott was retired and McClellan put in his place as general-in-chief. When Lincoln asked if perhaps he was overtaxing himself by retaining command of the Army of the Potomac while at the same time heading all the armies, McClellan replied, "I can do it all."[13]

Fall turned to winter, and the daily bulletins "All quiet on the Potomac" began to invite derisive laughter in Washington. Only the Federals were all quiet. The Rebels mounted batteries on the lower Potomac, closing the river to commercial traffic, and elaborately entrenched themselves around Manassas, just twenty-five miles from the capital. Stonewall Jackson prowled the upper Potomac above Washington, threatening the railroad and canal arteries. The capital of the United States was embarrassed under a partial blockade. Lincoln grew restive and set about putting pressure on General McClellan to do something—indeed, to do anything.

Silence was the tactic McClellan used to deflect unwanted advice, questions, or pressure. The historian George Bancroft, invited by Lincoln to army headquarters to meet McClellan, wrote of him afterward, "Of all silent, uncommunicative, reserved men, whom I ever met, the general stands among the first. He is one, who if he thinks deeply, keeps his thoughts to himself." Pennsylvania congressman Samuel S. Blair had a similar impression: "McClellan keeps his secrets well. I do not believe if he has a plan, that it is known to any one but himself." Certainly no plan was known to the commander-in-chief, and he was determined to find out if there was one.[14]

He did so by working out a plan of his own and presenting it to McClellan on or about December 1. Whether Lincoln discussed his idea beforehand with McClellan, or with any other military man, whether the original idea was his own or someone else's, is unknown. In any case, it was designed to challenge the Confederate army, now under Joe Johnston, threatening Washington. The president had studied the map and recognized the weakness of Johnston's position—his long and vulnerable supply line to Richmond via the Orange & Alexandria and Virginia Central railroads. A short turning movement by way of the line of the Occoquan River, within easy marching distance of the Washington lines, would threaten or cut the railroad south of Johnston's Manassas position, forcing him out of his entrenchments in order to defend his communications. Half the Federal army would make the turning movement while the other half made a feint against Manassas.

In outlining the Occoquan part of the operation, Mr. Lincoln suggested a theme he would return to again in his formulation of an offensive strategy. The movement there would be in two columns; if either column was resisted, the other might take the attackers in rear. "Both points will probably not be successfully resisted at the same time," Lincoln pointed out. It was an eminently practical plan, one nicely suited to circumstances. Coming as it did from the president, however, meant that for the sake of his self-respect General McClellan would have to go on record with a plan—a different plan—of his own devising.

In his response to the president's Occoquan plan, McClellan pointed out what to his mind was its fatal flaw. His intelligence information, he wrote, "leads me to believe that the enemy could meet us in front with equal forces *nearly*. . . ." What he meant by this awkward phrasing was that his sources counted the enemy at Manassas as 100,000, against a total force he could mobilize of 104,000. Thus at the objective point alone Johnston's strength nearly equalled McClellan's whole force, and it was known that Johnston had within easy supporting distance at least another 50,000 men. It went without saying that taking the offensive outnumbered three to two, and then dividing his own force in the face of such odds, would be most ill-advised. Then General McClellan suggested an alternative: "I have now my mind actively turned towards another plan of campaign that I do not think at all anticipated by the enemy nor by many of our own people."[15]

As 1861 turned to 1862, the president had to be content with this tantalizing hint of some new grand strategy. McClellan said nothing more about it, and then he fell ill with typhoid. Matters elsewhere were not progressing, either. Lincoln asked the army's quartermaster, Montgomery

Meigs, "General, what shall I do? The people are impatient; Chase has no money and tells me he can raise no more; the General of the Army has typhoid fever. The bottom is out of the tub. What shall I do?"

The president had been boning up on the military arts, borrowing General Halleck's *Elements of Military Art and Science* from the Library of Congress, finding out for himself conditions in the various theaters of war, and now he determined to act assertively as commander-in-chief by calling a council of war. On the evening of January 10, 1862, at the White House, he ordered in Generals McDowell and Franklin from the Army of the Potomac. From the administration came Secretaries Chase and Seward and Assistant Secretary of War Thomas A. Scott.

Lincoln began by recounting the discouraging news from every quarter; now McClellan's illness was casting an especial pall over the scene. He had been to McClellan's house but the general had not asked to see him. So far as he could see, nothing was being done anywhere—or would be done. By McDowell's account, Lincoln said that if General McClellan was not going to use the army, he wanted to borrow it, "provided he could see how it could be made to do something." He asked Generals McDowell and Franklin for their thoughts on the case.

McDowell's plan was much like Lincoln's of the month before: force Johnston out of his Manassas fortifications by threatening his communications. Franklin, a particular friend of McClellan's, had some hint of what was on the general-in-chief's mind and spoke of a scheme to move on Richmond by way of the lower Chesapeake and the York River. Lincoln invited them to study matters further and report back.

Word of the gathering reached McClellan through Secretary Chase, his particular patron in the cabinet, and he rose from his sickbed to attend a meeting of the council—the last meeting—on January 13. He put on an astonishing performance, sitting with his head down, mute, patently sulking. At the conference were three other generals—McDowell, Franklin, and Montgomery Meigs—and four leading figures in the administration—the president, Secretaries Seward, Chase, and Montgomery Blair—and they were far from being an unfriendly audience, yet McClellan sullenly refused every entreaty to make his case, to even hint at his plans in a way no more specific than his August 3 presentation to the cabinet.

Most astonishing of all was the reason he gave for this reticence. He confided to Meigs, who was urging him to respect the president's request, that if he revealed anything of his plans now, the next day they would be in the *New York Herald;* Lincoln could not even keep military secrets from Tad, his eight-year-old son. (On that very next day, in fact, McClellan unburdened himself in a three-hour briefing on every detail of the military situation to none other than a correspondent of the *New York Herald.* He said that what he had declined to tell the president he would now impart to editor James Gordon Bennett, "*all* the knowledge I possess myself, with no reserve. . . .")

George McClellan was marked by a streak of willful, self-destructive obstinacy, never more apparent than at this White House conference. Here was an ideal opportunity for him to bring everyone there into his camp as allies, without the slightest risk to the security of what would be described as his grand campaign. At that point, everyone in the room already knew at least its outlines. Instead, in the grip of what today would be described as nothing else but paranoia, he saw around him only enemies, and he alienated them all. As his patron Chase observed, "Well, if that is Mac's decision, he is a ruined man."[16]

The eight-week period beginning with this January 13 war council and extending into the first week of March 1862, proved to be the most intensely focused period of Lincoln's partnership with General McClellan. The president never let up on the pressure he was exerting on the general to get the war moving, and in many respects it was a productive time. At the beginning of the period Lincoln was complaining that the war effort was on dead center and the bottom was out of the tub. At the end of it, the war in the West and in key coastal areas had progressed nicely. In the East a strategic plan of campaign was in place and rapid progress made in preparing for its launch.

It was in this period, too, that Mr. Lincoln summarized his larger thoughts on military strategy, addressing Generals Buell and Halleck in the western theater. His letter elaborated a theme in his Occoquan plan and reflected a strikingly clear-cut common-sense view of the case. "I state my general idea of this war," he wrote, "to be that we have the *greater* numbers, and the enemy has the *greater* facility of concentrating forces upon points of collision; that we must fail, unless we can find some way of making *our* advantage an over-match for *his;* and that this can only be done by menacing him with superior forces at *different* points, at the *same* time; so that we can safely attack, one, or both, if he makes no change; and if he *weakens* one to *strengthen* the other, forbear to attack the strengthened one, but seize, and hold the weakened one, gaining so much."[17]

Although they began with wholly different premises, Lincoln and his general-in-chief reached similar conclusions regarding grand strategy. Lincoln was calling for pressure at many points from a reservoir of superior strength, looking for the weak point, the inevitable overmatch. The Young Napoleon's grand strategy also stressed coordinated movements by the Federals in the various theaters of war, but his purpose was to prevent the enemy from shifting forces so as to further add to the overmatch already facing the Army of the Potomac in the East. Both favored a movement into East Tennessee, for example, Lincoln to rescue the large Unionist element there, McClellan to cut the Confederacy's main lateral rail line and prevent reinforcements being thrown against him from the western theater.

A third party, Edwin M. Stanton, began playing a role in this period. Nominated for Cameron's post as war secretary on January 13, Stanton grew as determined as Lincoln that McClellan must be made to fight; as he put it, "the champagne and oysters on the Potomac must be stopped."[18] The new secretary of war was the prime mover—perhaps the originator—of a remarkable plan to shift 70,000 men and 250 guns from the Army of the Potomac to Kentucky to open a major winter campaign there. At first McClellan favored the scheme, particularly after he decided he would command it personally, but finally nothing came of it. The Kentucky movement was overtaken by events. The Fort Henry–Fort Donelson–Nashville operation began in Tennessee and moved swiftly toward a successful climax. Progress was made too on the North Carolina coast and in the build-up for the New Orleans campaign.

We have fewer details of how Lincoln dealt with General McClellan in this period than we would like, for a major source of information, McClellan's letters to his wife, is not available. Mrs. McClellan had come to Washington to be with her husband after his appointment as general-in-chief, and they remained together until his departure for the Peninsula on April 1. No doubt the general's personal life was happier that winter than at any time in his active service, but history is no wiser for that.

In any event, at some point in about mid-January, at Stanton's urging, McClellan discussed his plans verbally with the president. An undated Lincoln memorandum is surely from this discussion, and its contents—details of the Occoquan approach to Manassas—suggest that the president was determinedly pressing his plan on the general. How much McClellan revealed in turn of what would come to be known as his Urbanna plan is not known, but in light of Lincoln's next action, he cannot have been very forthcoming, or at least not very persuasive.

Lincoln's two war orders, dated January 27 and January 31, intended only for the eyes of General McClellan and the secretaries of war and navy, have been widely criticized by historians as intrusive interference in war operations. John Codman Ropes, writing in 1894, described the General War Order No. 1 of January 27 as "a curious specimen of puerile impatience." What is often overlooked, however, is the purpose behind these two orders (General War Order No. 1 specified "a general movement of the Land and Naval forces" to take place on February 22; Special War Order No. 1 of January 31 ordered the execution of the Occoquan plan). Since his appointment on November 1, General-in-Chief McClellan had only hinted at his strategic plans, and that rarely, or had flatly refused to divulge them even in the most general outline. It was true enough that Virginia was in the grip of its notorious mud season and that no general advance could now begin there before spring, yet to date no one in either the military or the civilian branch of the government (no one except General McClellan) knew if there was a single word on paper for what would prove to be the largest single military operation of the war. Mr. Lincoln's war orders did indeed signal his impatience, but there was nothing puerile about them. They served their purpose very nicely.[19]

When he received the president's Special War Order, ordering him to execute the Occoquan plan, on Friday, January 31, McClellan hurried to the White House to ask "his excellency" whether the order "was to be regarded as final, or whether I could be permitted to submit in writing my objections to his plan, and my reasons for preferring my own." We can be assured that his excellency had no real expectation that on Washington's Birthday every Federal soldier and sailor would advance in unison against the foe; what he sought, rather, and what it seemed he had now achieved, was to force General McClellan to reveal his hand.

McClellan worked all weekend on his plan. From a surviving draft in his papers we know he had started formulating his Urbanna scheme earlier in the month. Now his task was to elaborate that draft and put it in finished form. On Monday, February 3, his twenty-two-page plan was laid on Secretary Stanton's desk.

McClellan's Urbanna plan, which circumstances would modify to become his Peninsula plan, has inspired learned discussions of strategic turning movements and interior lines and exterior lines and the coupling of the strategic offensive with the tactical defensive. What is missing from this analysis, however, is McClellan's real purpose in proposing it. His idea of moving against Richmond from a new base on the lower Chesapeake was the sole way he could think of for the inferior army—his army—to defeat the considerably superior army of Joe Johnston. Every aspect of his planning, every movement of his forces over the next five months, was geared to what he took to be that reality.

McClellan selected Urbanna, a small tobacco port on the south bank of the lower Rappahannock, for its ideal location on the map. Federal seapower would be unchallenged carrying the grand army to this site just fifty miles from Richmond. One day's march, he said, would bring the army across to West Point on the York. From there two days' march along the Richmond & York River Railroad would bring the army to the gates of Richmond. Confederate forces on the Peninsula between the York and James would be cut off; the navy would guard his flanks and convoy his supplies to the railhead at West Point. At the heart of the plan was McClellan's expectation of gaining surprise, of reaching the Confederate capital before Johnston could get there from Manassas to defend it. Then any battle would be on ground of McClellan's choosing, with his army well entrenched and fighting defensively. Thus could the inferior army defeat the superior one.

McClellan chose interesting phrasing for this point. "The alternatives presented to the enemy," he wrote, "would be to beat us in a position selected by ourselves; disperse;—or pass beneath the Caudine Forks." His metaphor referred to the narrow defile in Italy in which a Roman army was trapped by Samnites in 321 B.C. and forced into abject surrender, bowing beneath a yoke of crossed spears. If everything went just right with the plan, General McClellan believed this to be a predictable outcome to his grand campaign.

Lincoln's continued defense of his Occoquan plan suggests that he did not think it necessary to go all the way to Richmond for an opportunity to defeat a Rebel army which just then was but twenty-five miles away. It further suggests that the president did not accept the disparity of forces that so influenced McClellan's thinking; he was willing to risk battle then and there with the army he had. Later, when McClellan was stalled before Yorktown on the Peninsula, Lincoln would remind him that he "always insisted" that going down the Chesapeake to new fields instead of engaging the enemy at or near Manassas was not surmounting the difficulty—"we would find the same enemy, and the same, or equal, intrenchments, at either place."

Yet however much he distrusted the Urbanna plan, the president had to accept it or force an unwanted plan on a general already inclined to drag his feet. Facing that truth, he yielded his plan to the general's. There is no small irony in the fact that as president Abraham Lincoln devoted far more time and effort to trying to influence the Peninsula campaign than to any other campaign of the war, and he never believed it was a campaign that needed to be fought.[20]

Whether or not George McClellan might have beaten Joe Johnston to Richmond by way of Urbanna and gained a bloodless victory thereby must remain a moot question. In the month after submitting his Urbanna paper McClellan collected forces and shipping for the movement, but offered no concrete planning and no timetable. A scheme he concocted to secure the upper Potomac and a position in the Shanandoah Valley became a fiasco when the canal boats needed for a floating bridge across the Potomac proved six inches too wide to pass through the canal lock into the river. The canal boat episode further tested the president's temper, and led him once again to find some prod to force McClellan into action.

Following what he described as a stormy private meeting on March 7 with the president, McClellan convened a council of war of his own and won a two-thirds majority of his generals in support of the Urbanna plan. The next day Lincoln announced his official approval, but stipulated there be no further delay. McClellan had ten days, until March 18, to begin the movement to this new base on the Chesapeake.

The president's war order to this effect was issued on March 8. On that day, and the next, the Confederates turned events upside down. The ironclad ram *Virginia* steamed out of Norfolk into Hampton Roads and devastated the Federal blockading squadron. Then Joe Johnston evacuated his Manassas lines for a new position to the south, behind the Rappahannock, and McClellan's scheme for basing his operations at Urbanna was gone beyond recall.

McClellan acted swiftly to salvage his grand campaign. To his mind, there was no real choice to be made. Should he recast his plan to operate overland toward Richmond from the line of the Rappahannock (as Burnside, Hooker, and U. S. Grant were to do), it would be a confession that the scheme he had defended so vehemently was untenable. Confession of failure was not in his make-up. He called together his generals for a second council of war and won approval for a revised plan, shifting the main base of operations from Urbanna on the Rappahannock to Union-held Fort Monroe at the tip of the James-York Peninsula. Richmond remained the objective by way of West Point and the railroad; only now West Point would be some thirty-five straight-line miles from the starting point. President Lincoln raised no objection to the change of base, but specified that the Manassas position be left well manned and that when the Army of the Potomac sailed off to the south it must leave Washington "secure."

Lincoln continued to be a most active commander-in-chief. Now that McClellan had "personally taken the field at the head of the Army of the Potomac," he relieved him of his duties as general-in-chief. The post was left vacant, suggesting that once Richmond was taken McClellan would resume its duties. Lincoln was equally active behind the scenes, defending McClellan from the attacks of the growing anti-McClellan cabal spearheaded by radical Republicans. The general, who was always well aware of political undercurrents, sensed the president's support and, for a rarity, appreciated it. On March 16 he wrote his New York mentor Samuel Barlow, "I shall soon leave here on the wing for Richmond—which you may be sure I will take," and he added, "*The President is all right*—he is my strongest friend."

That sentiment proved to be short-lived. Soon enough president and general parted company—and, on McClellan's part, never regained the moment of amity—over the question of the defense of Washington. Here was another occasion, like the war council meeting of January 13, when McClellan might have sat down with Lincoln and Stanton at the White House, carefully spelled out his plans for Washington's defense, sympathized with their concern for the safety of the city, and assured them that their concern was his concern too. Instead, on April 1 he sent to the War Department an incredibly carelessly drawn paper on the subject and set off for Fort Monroe, confiding to his wife that he was glad to be gone from the capital, "that sink of iniquity." He was apparently in another of his self-destructive moods.[21]

The Peninsula plan as it had now evolved was certainly a sound enough strategy for a general with McClellan's delusions about the enemy. It is hard to imagine him willingly taking the offensive against the Rebel host that spring in any other fashion. Fort Monroe was a perfectly secure main base, securely supplied by the navy. He also expected the navy to control both the James and York rivers, guard his flanks with gunboats, deliver his supplies, and permit him amphibious operations to turn enemy defenses. The Richmond & York River Railroad, running from West Point twenty-three miles straight into Richmond, offered him a fully adequate supply line to support his advance, and to carry the heavy siege guns he expected to need to offset his perceived inferiority in infantry. To be sure, he no longer thought of the Caudine Forks, of brilliantly maneuvering his foe into abject surrender. His thoughts instead were of the Siege of Sevastapol in the Crimea. He had made a thorough study of that operation, observing personally the siege lines there, and he looked with confidence toward a Siege of Richmond.

In his *Report,* published in 1864 after his relief from command, and in his memoir *McClellan's Own Story,* General McClellan would construct an elaborate explanation for the failure of the Peninsula campaign. He claimed the administration in Washington deliberately withheld needed reinforcements, tied him to an unprofitable line of advance against Richmond, and indeed that he and his army were victims of a conspiracy to see him defeated on the Peninsula and the war prolonged thereby until the abolitionists might gain the upper hand. While the editor of his posthumous memoir deleted Lincoln's name from McClellan's catalog of conspirators, from his first whiff of conspiracy McClellan had the president's name on his enemies list.

It must be said here that in any appraisal of George McClellan's view on any subject, the amount of truth in the view may be calculated in inverse proportion to the passage of time. The truth about his failure on the Peninsula, when drawn solely from contemporaneous sources, is not at all how McClellan drew it for history.

One example is the matter of the defense of Washington. McClellan's second war council, on Lincoln's order, had determined that 40,000 men should be assigned to the immediate defense of the capital. As late as March 30, McClellan himself assured Secretary Stanton that 50,000 men would "be left around Washington." When the general's April 1 paper on the subject was examined, however, it showed barely 26,700 "left around Washington," and most of those were raw recruits, untrained even in the use of their weapons. When Mr. Lincoln learned of this, said a White House visitor, "he was justly indignant."[22]

In consequence, McDowell's First Corps—three divisions—was held back to guard the capital. McClellan in his turn was indignant at the decision, and on it laid the blame for a sea of troubles. His plans were disrupted, his campaign stalled. In fact, he had already stalled his campaign by unnecessarily besieging Yorktown, and the First Corps had nothing to do with how the siege was conducted or how long it lasted.

His claim that he was unsupported is no more credible. He originally designed his campaign for 130,000 troops, and on the eve of its climax, in the Seven Days, he had received from the president 127,300, plus 7000 more under Burnside that were given to him to use if he could make up his mind what to do with them.

McClellan would claim, too, that he was restricted in his approach to Richmond by the need to link up with two First Corps divisions, sent him in due course as reinforcements, that Lincoln ordered to march to him overland. Otherwise, said the general, he would have taken the better James River route toward the Confederate capital. In truth, McClellan did not, either before or after the reinforcements reached him, ever contemplate shifting his campaign to the James. To do so would have meant abandoning the Richmond & York River Railroad, the only practical way to carry his siege guns close enough to Richmond to fight the enemy host on acceptable terms.

The fourth of McClellan's Peninsula myths was his assertion that restrictions placed on him by Lincoln prevented his winning the campaign by crossing the James and moving on Petersburg, as Grant would do in his final triumphant campaign in 1864–65. This suggested that the Young Napoleon might have shortened the war by two years had he been permitted his way.

It was General McClellan himself who in singular fashion ruled out any movement on Petersburg. He did so because of what may best be described as the Beauregard Bugaboo. This bugaboo haunted McClellan throughout the closing weeks of the Peninsula operation. Federals in the West had failed to cut the Confederates' direct rail route between Chattanooga and Virginia, and consequently McClellan feared that one day or the next Beauregard and his western army would fall on him. Should he turn the flank of the Rebels in Richmond by moving to Petersburg, he risked exposing his own flank to Beauregard's western troops. General Heintzelman attended a meeting on this subject, and that night wrote in his diary, "There will be no advance on Chattanooga by Buell for some time & in view of this Gen. McClellan opposes taking Petersburg."[23]

During these spring and summer months of 1862, Lincoln kept close to the Peninsula campaign through letters and telegrams. In the *Collected Works* are recorded fifty-three such military communications to General McClellan, including some of his most masterful. He and Stanton acted as their own general-in-chief, also giving direction to the campaign against Stonewall Jackson in the Shenandoah.

Nothing points up the difference in military thinking between president and general so dramatically as their respective views of Jackson's Valley campaign. When Jackson drove down the Valley as far north as the Potomac, Lincoln saw it as a unique opportunity to cut in behind him and trap him far from any support. He did his best to mobilize his Valley forces for such a coup; but for the limited abilities of his generals he might well have succeeded. The president also pointedly observed to General McClellan that every Confederate soldier counted in the Shenandoah Valley was one less soldier facing the Army of the Potomac before Richmond.

By contrast, McClellan could see no advantage for himself in the fact that, while Jackson might be holding in the Valley three Federal divisions slated for the Peninsula, by the same token these Federals just then were keeping Jackson's three divisions from joining the defenders of Richmond. By McClellan's reckoning, an enemy as rich in manpower as the Confederates could easily afford to maintain substantial forces in the Valley at no risk to themselves. No numbers remotely like those he was facing ever entered his calculations.

Periodically Mr. Lincoln would make a carefully considered effort to reason with the Young Napoleon, who as he inched his way up the Peninsula was acting very un-Napoleonic. When the general complained, for example, that he was stymied at Yorktown and outmanned by the enemy and unsupported by the government, Lincoln sent him a letter that patiently explained why he had felt obliged to retain McDowell's corps at Washington, aptly compared the former situation at Manassas to the present one in front of Yorktown, and sought to bring enlightenment to the general on the larger situation. "And, once more let me tell you, it is indispensable to *you* that you strike a blow. *I* am powerless to help this. . . . I beg to assure you that I have never written you, or spoken to you, in greater kindness of feeling than now, nor with a fuller purpose to sustain you, so far as in my most anxious judgment, I consistently can. *But you must act.*" McClellan was unmoved by the reasoning, and the reasonableness. After the president pointed out to him that time was his enemy and he should move against Yorktown quickly, McClellan reported this to his wife and added, "I was much tempted to reply that he had better come & do it himself."[24]

That might well have yielded results. Early in May the president did go to the Peninsula and promptly brought about an important advance in the campaign. He ordered and organized the capture of Norfolk by the Fort Monroe garrison and in so doing hastened the fate of C.S.S. *Virginia.* For the past month the Rebel ironclad had seriously crimped McClellan's operations by blocking access to the James, but now the capture of her Norfolk base left no time to lighten ship sufficiently to withdraw far up the James to a new blocking position at Harrison's Landing. In that position she would have truly crimped McClellan's operations. Instead, on May 11, the *Virginia* was blown up by her crew.

Despite all the things that went wrong in executing his Peninsula plan, General McClellan almost surely would have succeeded in putting Richmond under siege and in due course capturing it but for the Yankee shot and shell at Seven Pines that elevated Robert E. Lee to command. "The shot that struck me down is the very best that has been fired for the Southern cause yet," Joe Johnston observed, referring to Richmond's confidence in the new commander of the Army of Northern Virginia.[25] The real truth of his remark was that now the South had a general in command who understood exactly how to beat George B. McClellan, and in the Seven Days' Battles proceeded to do so. It was not a polished performance—like all Civil War generals, Robert E. Lee had to learn his trade while practicing it—but it was highly effective all the same.

With the Army of the Potomac driven back from the gates of Richmond to Harrison's Landing on the James, President Lincoln made a second trip to the Peninsula to see the case for himself. His visit marked a turning point in the partnership of president and general. Perhaps this was due in part to the famous Harrison's Landing letter that McClellan handed him on July 8. The president can hardly have been surprised by this lecture on how to manage the war, or by McClellan's judgment of the slavery question. The "forcible abolition of slavery," the general wrote, must not be contemplated "for a moment," and he warned, "A declaration of radical views, especially upon slavery, will rapidly disintegrate our present Armies." The only record of Lincoln's reaction to the letter is in McClellan's account, which has the president reading it, thanking him, and saying nothing further about it. This disappointed the general. He told his wife, "His Excellency . . . really seems quite incapable of rising to the heights of the merits of the question & magnitude of the crisis."[26]

For his part, Lincoln surveyed the ruins of the Peninsula campaign and concluded that there was nothing indispensable about General McClellan. Already, in the midst of the campaign, he had given momentary thought to replacing him, and not long after his return to Washington he would try (and fail) to persuade Burnside to take the command. He also called in Henry Halleck from the West to be his new general-in-chief, telling him he might replace the commander of the Potomac army if he liked. Clearly, George McClellan had failed the test the president was now applying to all his generals—did they know how to win on the battlefield?

With Halleck in place as general-in-chief, the president deliberately distanced himself from the commander of his principal army. In the seven weeks after his visit to Harrison's Landing, he sent but four communications to McClellan, only one of which was of any consequence. (What had happened, he wanted to know, to the 20,000 to 30,000 soldiers who had simply disappeared from the army during the late campaign?) While he kept a watch on the developing Second Manassas situation, Lincoln left it to General Halleck to coordinate the melding of McClellan's Army of the Potomac with Pope's Army of Virginia.

Flares of temper by Abraham Lincoln not infrequently involved General McClellan. One such occurred on August 29 when in the midst of the Second Manassas crisis McClellan telegraphed to the president that the course of action he favored was "To leave Pope to get out of his scrape & at once use all our means to make the Capital perfectly safe." A newspaperman at the White House would report that he had never seen the president "so wrathful as last night against George." Mr. Lincoln, recorded John Hay in his diary, "said it really seemed to him that McC. wanted Pope defeated." This stated the case a bit too strongly—too many of McClellan's own troops were serving under Pope for him to wish that—yet McClellan fully expected Pope to be defeated and thought he deserved to be.[27]

It was in this context that the president took what was surely for him the most difficult military-command decision he made during the war. When he returned McClellan to command—more properly, when he returned the Army of the Potomac to McClellan—on September 2, following Pope's defeat, he discussed it with no one but Halleck. Indeed, it was counter to the wishes of the majority in his cabinet, who intended that day to present him with a petition calling for the general's dismissal. By Attorney General Bates's account, Lincoln "was in deep distress . . . seemed wrung by the bitterest anguish." Gideon Welles reported his saying that McClellan "can be trusted to act on the defensive, but having the 'slows' he is good for nothing for an onward movement." Still, the decision was as inevitable, and as logical, as the decision the previous July that brought McClellan to Washington. Lincoln put it in clear perspective: "McClellan has the army with him."

Just then that was the paramount issue. In Washington there was genuine fear that the troops might not fight if called on again by Pope and McDowell. Whatever his failings, there was no doubt that the men would fight for McClellan. When it was reported that the Confederates were on the march again, Lincoln had to make a second highly difficult decision—to permit McClellan to take the field in pursuit. He did so on September 3, during a visit to McClellan's Washington house in company with General Halleck.

Interestingly, the president did not take responsibility for this second decision, but instead said it was Halleck who made it; he supported him (against his better judgment, he said) so as not to undercut the general-in-chief's authority.

"I could not have done it," he told Welles in speaking of the decision, "for I can never feel confident that he will do anything." For his part, Halleck said that it was the president who put McClellan in field command, that he knew nothing of it beforehand. On his record of indecisiveness during this period, it is improbable that Halleck would have taken a major decision like this on his own. Like the decision the previous day, it was the president who made it.

Lincoln's reason for denying accountability was no doubt political. Even the loudest among the anti-McClellan cabal had to admit the general was competent enough to defend Washington against an assault. But commanding in the field, during an "onward movement," was another matter, and if in the coming days the army under General McClellan should be defeated, Mr. Lincoln was not anxious to have it known that it was he who had sent the general into the field. By way of confirming his discomfort at the decision, after leaving McClellan's house he tried again—and failed again—to persuade General Burnside to take the command. "We must use what tools we have," he said resignedly to John Hay.[28]

So it may be assumed that the president followed the progress of the Maryland campaign with fingers figuratively crossed. As he had with Jackson in the Valley in the spring, Lincoln saw this march north by the Confederates as an opportunity to get in their rear and cut their communications and bring them to battle on favorable terms. General McClellan, however, appeared to be making no vigorous effort in this direction, and instead turned to his litany of being outnumbered. The general finally settled on a count of 120,000 for Lee's army, three times its actual strength.

In the end, it was the Yankee soldiers of the Army of the Potomac who ratified Lincoln's decision to turn again to General McClellan by fighting for their general, and fighting magnificently, at Antietam. Rather than engaging in battle on his own terms, as he expected to do, Lee had to engage on McClellan's terms, thanks to the Lost Order. Antietam was one of the very rare Civil War battles in which the war might have been decided in an afternoon. That it was not was due in equal measure to Lee's battlefield brilliance and McClellan's battlefield timidity.

Yet after this bloody stand-off on September 17th, Lee was forced to return to Virginia, and the war moved on to a different direction. If it was not the victory Lincoln had hoped for, it was for him victory enough to issue the preliminary Emancipation Proclamation. The general who had warned that emancipation would disintegrate the armies had headed the army that made emancipation possible. McClellan was appalled at this turn of events, and at Lincoln's concurrent suspension of the habeas corpus privilege.

These presidential actions seemed to General McClellan to extinguish his last hope for restraining the war within decent bounds. He could no longer say, as he had in the war's early days, that he understood and shared "your Excellency's . . . course & opinions. . . ." Lincoln's "national strategy," in James McPherson's phrase, had now emerged in a form wholly alien to George McClellan.

McClellan always viewed secession as an aberration, foisted on the good people of the South by rabble-rousers and fire-eaters. He believed that once their army was defeated and their capital taken and their puppet government deposed, in warfare conducted "upon the highest principles known to Christian Civilization" (as he put it in the Harrison's Landing letter), Southerners would come to their senses and welcome negotiations. As the Young Napoleon imagined it, he as conqueror would preside over a peace conference where, after certain adjustments were agreed to and certain concessions made regarding the South's inalienable rights, Southerners would sign a treaty renouncing secession and pledging their national allegiance.

The general was realist enough to know that at the peace table the slavery issue would have to be faced. "When the day of adjustment comes," he told his wife, "I will . . . throw my sword into the scale to force an improvement in the condition of those poor blacks." While he would never throw his sword into a fight for abolition, he would support any gradual emancipation that recognized equally the rights of master and chattel.

Indeed, at a flag of truce meeting on prisoner exchange during the Peninsula campaign, McClellan had broached just such a negotiated peace with Howell Cobb, onetime member of Buchanan's cabinet and now a brigadier in the Rebel army. The North held no thought of subjugation, Cobb was assured; it sought instead only to uphold the Constitution and to enforce the laws in all the states. Slavery, recently excluded by law from federal territories and abolished in the District of Columbia, need no longer be a divisive issue. Only after Cobb rejected this peace feeler did McClellan bother to inform Washington of the proceedings.

The actions of the president and the government had steadily eroded McClellan's hope for a war of restraint. He was deeply disturbed by congressional passage of the first and second confiscation acts, and by what he took to be the draconian measures inflicted on Southern civilians during

General Pope's tenure as army commander. Such things, he believed, would demoralize the troops and lead to pillage and rapine. On the Peninsula he had genuinely feared for Richmond's citizens after a successful siege, as if his men were sixteenth-century soldiery promising the sack of some captured walled city in the Netherlands.

Now, in September 1862, came the last straw: Lincoln embracing abolitionism and overriding the constitutional protection of habeas corpus. To one of his homefront supporters General McClellan spoke of "the recent Proclamations of the Presdt inaugurating servile war, emancipating the slaves, & at one stroke of the pen changing our free institutions into a despotism," and for a time he seriously considered taking a public stand against the "accursed doctrine" of emancipation.

In the general's mind, all these actions smacked of radicalism, his catch-all term for whatever disturbed or overturned conservatism in all things political or military. In an article suggesting the Union's proper military strategy, published anonymously in 1864, he would urge that Washington's policy toward the South "be in accordance with the enlightened maxims of the New Testament, not with the bloody and barbarous code of the nations of old time, who fought solely to destroy and enslave." In the same vein, McClellan would see it as his duty to oppose Lincoln in the election of 1864. As he put it upon entering the political arena, the "sole great objects of this war are the restoration of the unity of the nation, the preservation of the Constitution, and the supremacy of the laws of the country."[29]

To George McClellan, at least, it had become clear that he and the president were growing far apart on a national strategy with which to wage the war, and that furthermore they were on diverging paths. This realization helped shape his conduct in the weeks after Antietam, and close study of his letters, especially his private letters, suggests that he was not greatly surprised when Lincoln relieved him of command on November 7.

Indeed, McClellan had built such as case for his honorable retirement from field command that in his most secret of hearts being relieved may have come almost as a relief. His linkage of the dismissal of both Stanton and Halleck to his continuation in command offered him the excuse (and the rationale) he always desired for major changes in his life. Hints in his letters to his wife also suggest he was not anxious to renew confrontation against Robert E. Lee. "I feel that the short campaign just terminated will vindicate my professional honor & I have seen enough of public life," he told her. "No motive of ambition can now retain me in the service. . . ."

Lincoln did not reach his decision lightly. The public's perception of Antietam depended a good deal on which newspaper one read and the coloration of one's politics, yet clearly the battle was a good deal closer to a victory than to a defeat, and General McClellan could at the very least be said to have ended the Confederate invasion of the North. Lincoln seems to have sensed the opportunity for a great victory in Maryland that McClellan mishandled, but missing a great victory was a hard case to make as a reason for dismissal.

Lincoln visited the army in Maryland early in October and met alone with his general. He apparently went to some lengths to reason with him and to discuss frankly what he called his overcaution and to try and persuade him to renew the campaign. "The Presdt was very kind personally—told me he was convinced I was the best general in the country, etc etc.," McClellan wrote his wife. "He was very affable & I really think he does feel very kindly towards me personally." He promised to give thought to a new campaign but volunteered nothing beyond that. It was on this visit that Mr. Lincoln made his famous remark to his friend Ozias Hatch that the army might be called the Army of the Potomac, "but that is a mistake; it is only McClellan's bodyguard."[30]

Some ten days after his meeting with McClellan in Maryland, the president sat down and composed the best of all his letters to this general; arguably it is the best letter he wrote to any of his generals. It is worth quoting at length:

> You remember my speaking to you of what I called your over-cautiousness. Are you not over-cautious when you assume that you can not do what the enemy is constantly doing? Should you not claim to be least his equal in prowess, and act upon the claim?
>
> As I understand, you telegraph Gen. Halleck that you can not subsist your army at Winchester unless the Railroad from Harper's Ferry to that point be put in working order. But the enemy does now subsist his army at Winchester at a distance nearly twice as great from railroad transportation as you would have to do without the railroad last named. He now wagons from Culpepper C.H. which is just about twice as far as you would have to do from Harper's Ferry. He is certainly not more than half as well provided with wagons as you are. I certainly should be pleased for you to have the advantage of the Railroad from Harper's Ferry to Winchester, but it wastes all the remainder of autumn to give it to you; and, in fact

ignores the question of *time,* which can not, and must not be ignored.

Again, one of the standard maxims of war, as you know, is "to operate upon the enemy's communications as much as possible without exposing your own." You seem to act as if this applies *against* you, but can not apply in your *favor*. Change positions with the enemy, and think you not he would break your communication with Richmond within the next twentyfour hours? You dread his going into Pennsylvania. But if he does so in full force, he gives up his communications to you absolutely, and you have nothing to do but to follow, and ruin him; if he does so with less than full force, fall upon, and beat what is left behind all the easier.

Exclusive of the water line, you are now nearer Richmond than the enemy is by the route that you *can,* and he *must* take. Why can you not reach there before him, unless you admit that he is more than your equal on a march. His route is the arc of a circle, while yours is the chord. The roads are as good on yours as on his.

You know I desired, but did not order, you to cross the Potomac below, instead of above the Shenandoah and Blue Ridge. My idea was that this would at once menace the enemies' communications, which I would seize if he would permit. If he should move Northward I would follow him closely, holding his communications. If he should prevent our seizing his communications, and move towards Richmond, I would press closely to him, fight him if a favorable opportunity should present, and, at least, try to beat him to Richmond on the inside track. I say "try"; if we never try, we shall never succeed. If he make a stand at Winchester moving neither North or South, I would fight him there, on the idea that if we can not beat him when he bears the wastage of coming to us, we never can when we bear the wastage of going to him. This proposition is a simple truth, and is too important to be lost sight of for a moment. In coming to us, he tenders us an advantage which we should not waive. We should not operate as to merely drive him away. As we must beat him somewhere, or fail finally, we can do it, if at all, easier near to us, than far away. If we can not beat the enemy where he now is, we never can, he again being within the entrenchments of Richmond . . .

I should think it preferable to take the route nearest the enemy, disabling him to make important move without your knowledge, and compelling him to keep his forces together, for dread of you. The gaps would enable you to attack if you should wish. For a great part of the way, you would be practically between the enemy and both Washington and Richmond, enabling us to spare you the greatest number of troops from here. When at length, running for Richmond ahead of him enables him to move this way; if he does so, turn and attack in rear. But I think he should be engaged long before such point is reached. It is all easy if our troops march as well as the enemy; and it is unmanly to say they can not do it.

This letter is in no sense an order. Yours truly

A. LINCOLN.

This was Abraham Lincoln at his best, full of common sense and sound military judgments; his image of the two armies advancing through Virginia on the arc and the chord of a circle is remarkably apt. He expressed as well something he felt deeply about—his belief that northerners were as good soldiers as those of the enemy, that they were just as capable of doing "what the enemy is constantly doing." At the same time, there is here a sense of the last resort, of one final and carefully crafted effort to reach his general and make him see reason.

McClellan replied, "I promise you that I will give to your views the fullest & most unprejudiced consideration, & that it is my intention to advance the moment my men are shod and my cavalry are sufficiently remounted to be servicable." Privately he told one of his lieutenants, "Lincoln is down on me," and predicted that he would be relieved and sent to a command in the western theater.[31]

At last, following a direct order from Washington, the army began crossing the Potomac into Virginia. The crossing alone took nine days, and beyond resuming the familiar tale that he was substantially outnumbered, McClellan said nothing of his plans. With great deliberation he moved along the chord of Lincoln's circle, while the Confederates moved swiftly on the arc, and by November 4 it was known at the White House that the enemy was blocking McClellan's path to Richmond. That same day the last of the midterm elections was held in the North, and that too entered into Lincoln's calculations regarding this most political of generals. The next day he sent the directive to Halleck to relieve General McClellan.

For nineteen months, until almost the end of 1862, the president had spent uncounted hours with McClellan and uncounted more hours in dealings about him; no other field commander throughout the course of the war required of him anything comparable in time and effort. Lincoln must be credited with being the moving force behind implementing the Peninsula campaign—a campaign he never approved of. He had to take the wrenching decision to restore McClellan to command during the Second Manassas crisis, and finally to dismiss him from all command barely two months later.

Theirs should have been a successful partnership if measured by the effort expended on it by this greatest of American presidents. In the end, however, Lincoln could do everything for this general but make him fight—and in the end, that is the measure of the general. George McClellan was simply in the wrong profession. He might have been highly successful in many fields, whatever his inclination, but field command in a civil war was not one of them.

When Alexander McClure was about to publish his *Abraham Lincoln and Men of War-Times,* in 1892, he wrote to McClellan's widow in advance explanation of the chapter in the book he titled "Lincoln and McClellan." In summing up his own view of the two men for Mrs. McClellan, he in effect composed an epitaph for their partnership. "The General's single mistake, that was the source of all his misfortunes, was his distrust of Lincoln," McClure said. "Had he understood and treated Lincoln as his friend, as I knew Lincoln was, he could have mastered all his combined enemies."[32]

2

Wilderness and the Cult of Manliness: Hooker, Lincoln, and Defeat

Mark E. Neely, Jr.

LINCOLN AND HOOKER ON DRESS PARADE

President and Commanding General inspect the Army of the Potomac. "The event of the season," the *New York Herald* saw fit to describe the show in the spring of 1863, on the eve of the Battle of Chancellorsville. *(Detail from the wood engraving of Alfred R. Waud that appeared in* Harper's Weekly. *Boritt Collection)*

JOSEPH HOOKER AND ULYSSES S. GRANT would have looked remarkably alike to a president reviewing their records early in 1863. They had their flaws, but both were judged to be among the best generals in the Union army. Both gained their promotions on merit. Both were reasonably successful and consistently hard-fighting professional soldiers with about as much experience of warfare as an American could get. By early 1863 each had engaged in the heaviest fighting of the war, Grant at Shiloh and Hooker at Antietam.

Both Hooker and Grant were West Point graduates of middling class rank. Each served with distinction in the Mexican War. Both subsequently spent time in California. Both had resigned from the peacetime army and failed conspicuously in civilian pursuits. Apparently one of the reasons Henry W. Halleck did not like Hooker was that Hooker had borrowed money from him in California and never paid it back.[1] When war broke out in 1861, Hooker did not have enough money to purchase a ticket to go east to Washington to seek a commission in the army, and friends had to subsidize his journey across the continent. Rumors of serious drinking problems swirled around both Hooker and Grant.

There were some differences, of course. Grant was a good family man who loved his wife and never swore. Hooker was a bachelor and a tough talker, a braggart in fact. More important in terms of high command, talking was a serious problem with Hooker, who had a reputation for undermining his superiors in the chain of command by outspoken and fractious criticism. Grant tended to suffer quietly. But who can blame Hooker for attempting to halt Ambrose Burnside's pathetic assaults on Fredericksburg or for criticizing that miserably failed commander afterward?

In the end, Grant became a conspicuous success as a commander and Hooker, a conspicuous failure. But there was no way early in 1863 to predict their future careers. Lincoln, of course, did not choose between Hooker and Grant in 1863 or at any other time. The point of this comparison is simply to suggest that it is difficult to choose generals to command independent armies.

Under the American system, civilian control of the military—ultimately control by politicians—is so complete that it is always tempting to think that political considerations necessarily influenced and perhaps hampered making wisely objective choices of military commanders. Because President Lincoln wrote a letter to Hooker soon after his appointment that in later years became quite famous, an atmosphere of controversy and political mystery have surrounded the president's decision to appoint him for almost a century. Written January 26, 1863, the confidential letter stated:

> I have placed you at the head of the Army of the Potomac. Of course I have done this upon what appear to me to be sufficient reasons. And yet I think it best for you to know that there are some things in regard to which, I am not quite satisfied with you. I believe you to be a brave and a skilful soldier, which, of course, I like. I also believe you do not mix politics with your profession, in which you are right. You have confidence in yourself, which is a valuable, if not an indispensable quality. You are ambitious, which, within reasonable bounds, does good rather than harm. But I think that during Gen. Burnside's command of the Army, you have taken counsel of your ambition, and thwarted him as much as you could, in which you did a great wrong to the country, and to a most meritorious and honorable brother officer. I have heard, in such way as to believe it, of your recently saying that both the Army and the Government needed a Dictator. Of course it was not *for* this, but in spite of it, that I have given you the command. Only those generals who gain successes, can set up dictators. What I now ask of you is military success, and I will risk the dictatorship. . . .
>
> And now, beware of rashness. Beware of rashness, but with energy, and sleepless vigilance, go forward, and give us victories.

There was nothing political about Hooker's appointment. Hooker's biographer was not able to determine clearly whether Hooker was a Democrat or a Republican during the Civil War. More important, neither could the public in Lincoln's day. The appointment sparked little controversy. In the month following its announcement, the president did not receive one single letter about it, pro or con. Republicans did not praise it nor Democrats condemn. No one wrote the president to warn him that Hooker was scheming to put a dictator in the White House.[2]

To be sure, Hooker did have the problem of *whispered* allegations that he advocated a dictatorship. And in that way he was quite different from Grant. Ulysses S. Grant scrupulously avoided such political themes and always gave sound answers on political questions when asked. For that matter, the president was as much at fault for his own credulity and unfounded political anxiety as Hooker was for loose and dangerous talk. Lincoln proved willing in this instance to call Hooker on the carpet for talking of dictatorship on the strength of a hearsay remark made, not in a private appointment in the president's office but at a social reception in the public rooms of the White House, a remark from a newspaper editor who had heard of Hooker's loose talk not firsthand but from one of his reporters.[3] Civil War journalism lacked the most rudimentary professional standards, and newspaper scuttlebutt deserved even less respect then than now. In other words, the Hooker letter proved that Abraham Lincoln feared dictators as much as it proved that Joseph Hooker wanted to be one.

Lincoln chose him because on paper Hooker looked like the best bet to bring victories to the hapless Army of the Potomac. The president consulted no one, neither politician nor general, in the decision. He did not have time to. Ambrose Burnside appeared on Lincoln's doorstep at six in the morning of January 25, 1863, threatening resignation, and by ten o'clock Lincoln had decided on Hooker as a replacement.[4]

The selection of Joseph Hooker was consistent with Lincoln's customary method of exercising his powers as commander-in-chief (he took the duties seriously and fairly literally), and the general was as good a risk as any. Another way to appreciate the difficulty in choosing generals is to compare Lincoln's performance with that of other contemporary heads of state. Even where military institutions were more structured than in the United States, the choices always proved problematic. British experience in the same era was so dismal that one military historian concluded that "whatever other aptitudes Victorians may have possessed, military proficiency was clearly lacking." Great Britain in the Crimean War, some five years before the American Civil War, produced command decisions that were and still are a byword and symbol of military folly: "The Charge of the Light Brigade" conjures up the very image of military unreasonableness. The British were especially handicapped by the system of having officers purchase their commissions.[5]

The French fared better in the Crimea—and that is why the Americans in the Civil War wore kepi hats and formed Zouave regiments in imitation of them—but Napoleon III's best command decision was not to go into the field himself. He was thinking about it, but advisers dissuaded him for fear there would be a revolution at home if he left the country for the faraway Crimea. The British did not want him to go either. Five years after the American Civil War, in the Franco-Prussian War, Napoleon III was more under the influence of the empress, his wife Eugénie, and she wanted him to take command in the field. He did, and the result was the squandering of what was widely regarded as the finest army on earth.[6]

As for the victors in that war, so famous for their military prowess, the Prussians, their officer corps might be described as narrowly restricted to a hidebound aristocracy. Even with a militaristic constitution, frequent wars for practice, the unique institution of a general staff to study and plan for war all the time, and Helmuth von Moltke as de facto commander-in-chief, selection of commanders could not be made from military considerations alone. Moltke sneered at the amateurism of the American conflict—it was not so much warfare as "the movement of armed mobs"—but five years later he chose as commanders of the three great German armies in the Franco-Prussian War men whose professionalism was not their only recommendation, the king's son, the king's nephew, and a seventy-four-year-old born in the eighteenth century, who, in the words of historian Michael Howard, "brought the whole German army within measurable distance of catastrophe."[7]

By standards of comparison with his international peers, Lincoln performed reasonably well. If Hooker too brought his army "within measurable distance of catastrophe," there was no predicting his poor performance in January 1863. Twentieth-century military historians fault him most for losing the initiative—for maneuvering successfully onto Robert E. Lee's flank only to surrender the offensive and pull back to a defensive position. How could Lincoln have suspected that Hooker would fail in this manner—through loss of confidence? In fact the fears of the time ran quite the opposite way. "Beware of rashness," Lincoln cautioned him in the famous letter of January 26. But to lose one's confidence upon assuming independent command was a common military failing. As far back in American history as the Revolution, General Lafayette, wise beyond his twenty-four years, commented on his own advent to independent command in Virginia: "To speak truth, I become timid in the same proportion as I become independent. Had a superior officer been here, I could have proposed half a dozen of schemes."[8] Such caution or paralysis in the face of the enemy when in ultimate command was an inevitable risk of promoting someone to independent command. President Lincoln—even Joseph Hooker himself—had to wait to see what would happen.

Despite Hooker, Chancellorsville, and other defeats, Abraham Lincoln usually receives high marks as a commander-in-chief. Since World War I, especially, historians have praised him for two related virtues: first, belief in attack, and second, courage or belief in himself.[9] These usually conspired to cause Lincoln to urge his generals, as he did Hooker in the famous letter, to "go forward, and give us victories."[10] Indeed, we know that Lincoln had reached a profound understanding of nineteenth-century warfare on the eve of Hooker's appointment to command in the East from remarks he made to one of his private secretaries in the White House after Burnside's defeat at Fredericksburg in December 1862. The secretary recalled:

> We lost fifty per cent more men than did the enemy, and yet there is sense in the awful arithmetic propounded by Mr. Lincoln. He says that if the same battle were to be fought over again, every day, through a week of days, with the same

> relative results, the army under Lee would be wiped out to its last man, the Army of the Potomac would still be a mighty host, the war would be over, the Confederacy gone, and peace would be won at a smaller cost of life than it will be if the week of lost battles must be dragged out through yet another year of camps and marches, and of deaths in hospitals rather than upon the field. No general yet found can face the arithmetic, but the end of the war will be at hand when he shall be discovered.[11]

I describe this, anachronistically, as understanding the nature of war before the germ theory of disease, and it was substantially the view of warfare held by the general who closed out the war for the North, Ulysses S. Grant. Horace Porter, who served on Grant's staff, described the general's reasoning for the assault on Cold Harbor this way (note how much it sounds like Lincoln):

> The general considered the question not only from a military standpoint, but he took a still broader view of the situation. The expenses of the war had reached nearly four million dollars a day. Many of the people in the North were becoming discouraged at the prolongation of the contest. If the army were transferred south of the James without fighting a battle on the north side, people would be impatient at the prospect of an apparently indefinite continuation of operations; and as the sickly season of summer was approaching, the deaths from disease among the troops meanwhile would be greater than any possible loss encountered in the contemplated attack.[12]

As for personal courage, Lincoln was a modest man and therefore did not think he owned this virtue in any abundance, but he came to realize that he possessed more of it than many military men. "Often," the president told another of his private secretaries on April 28, 1864, "I who am not a specially brave man have had to sustain the sinking courage of these professional fighters in critical times."[13]

Unfortunately, Lincoln's virtues as commander-in-chief were not virtues in every situation, and the Chancellorsville campaign presented precisely such a situation. Put simply, the problem was terrain. Lincoln's virtues as commander-in-chief were useless in a wilderness.

Terrain is the most underestimated factor in the military history of the Civil War. Conversely, the importance of terrain is the clearest lesson of even the most modern wars. Given jungle or mountain terrain, a tiny underdeveloped third-world nation like Vietnam can hold a superpower at bay for years. Eliminate that unfavorable terrain as a factor by turning it into level desert, and the superpower will bury the third-world nation in a matter of weeks, as the United States and it allies did Iraq in the Gulf War in 1991. Unhappily for Jefferson Davis, the Confederacy was not everywhere protected by mountain ranges, and what help the mountains offered him physically was usually undermined by a political problem, the disaffection of the subsistence farmers of the Confederate uplands, who were not fond of the Confederacy's ruling elite of slaveowners from the lowlands.[14] There were no jungles in the Confederacy, either. But the Wilderness of northern Virginia came close. And when Union armies fought there, they lost.

In truth, the factor of terrain was commonly overlooked by military commentators at the time. It required the sharp engineer's eye of Gouverneur Warren, for example, to recognize the problem in Virginia. After Chancellorsville, Warren submitted a remarkable report, which included the following observations:

> All our known topography in the entire region from the Potomac to the James River, and from the Blue Ridge to the Chesapeake . . . is a dense forest of oak or pine, with occasional clearings, rarely extensive enough to prevent the riflemen concealed in one border from shooting across to the other side; a forest which, with but few exceptions, required the axmen to precede the artillery from the slashings in front of the fortifications of Washington to those of Richmond. . . . It will aid those seeking to understand why the numerous bloody battles fought between the armies of the Union and of the Secessionists should have been so indecisive. A proper understanding of the country, too, will help to relieve the Americans from the charge so frequently made at home and abroad of want of generalship in handling troops in battle—battles that had to be fought out hand to hand in forests, where artillery and cavalry could play no part; where the troops could not be seen by those controlling their movements; where the echoes and reverberations of sound from tree to tree were enough to apall the strongest hearts engaged, and yet the noise would often scarcely be heard beyond the immediate scene of strife.[15]

General Warren thus supplied one of the finest introductions in the English language to the war in Virginia. This report, and not the now customary litany about technology—railroads, rifled firearms, rifled artillery, and so forth—should be the required introduction to any history of the Civil War, at least of the eastern theater.[16]

Warren's more detailed description of the Chancellorsville area likewise superbly set the scene for Hooker's campaign in 1863:

> At the time the operations resulting in the battle of Chancellorsville and those attending it began, the enemy occupied in strong force the heights south of the Rappahannock River, from Skinker's Neck to Banks' Ford, having continuous lines of infantry parapets throughout (a distance of about 20 miles), his troops being so disposed as to be readily concentrated on any threatened point. . . . Stafford County, in which the Union Army was located, is noted for its poverty. A lack of fertility in the soil has discouraged enterprise, and the country is wanting in public improvement such as are usually to be found in more prosperous communities. Dense woods and thickets of black-jack oak and pine cover most of the ground. The general character of the country is that of a wilderness, and it forms part of that distinguishing belt of country which continues through Orange and Spotsylvania Counties and southwesterly in a general direction parallel with the Blue Ridge. . . . Spotsylvania County, on the south side of the Rappahannock, in which Chancellorsville . . . is situated, is much of the character of Stafford, just described. . . . The term "Wilderness" is localized in common parlance for a portion of this country, and no one can conceive a more unfavorable field for the movements of a grand army than it presents. . . .[17]

Such terrain, as we in the United States know from modern and bitter experience, can defeat superior numbers and the most advanced technology.

To emphasize attack in such an environment was not a

virtue. To be sure, Robert E. Lee and the Confederate army operated in the same environment, but as Lincoln knew already, they enjoyed the advantages of interior lines and of superior intelligence. As Lincoln put it in his famous letter to Count Gasparin on August 4, 1862, explaining the advantages enjoyed by the Confederacy that compensated for their inferiority in numbers and resources, "The enemy holds the interior, and we the exterior lines; and . . . we operate where the people convey information to the enemy, while he operates where they convey none to us." Moreover, Lincoln's early realization of the nature of warfare before the germ theory of disease came to him from a habit of thought that did not always lead to sound military advice. Because he had risen to success from a hard youth spent as an ambitious boy in a lazy family, Lincoln despised avoiding obstacles instead of confronting them. When back in 1851 his worthless step-brother had considered moving from Illinois to Missouri to get another start in life, Lincoln wrote him sound but somewhat cruel advice:

> I . . . can not but think such a notion is utterly foolish. What can you do in Missouri, better than here? Is the land any richer? Can you there, any more than here, raise corn, & wheat & oats, without work? Will any body there, any more than here, do your work for you? If you intend to go to work, there is no better place than right where you are; if you do not intend to go to work, you can not get along any where. Squirming & crawling about from place to place can do no good.

Eleven years later President Lincoln proffered the same kind of advice on April 9, 1862, to his squirming and crawling commander of the Army of the Potomac, George B. McClellan:

> It is indispensable to *you* that you strike a blow. *I* am powerless to help this. You will do me the justice to remember I always insisted, that going down the [Chesapeake] Bay in search of a field, instead of fighting at or near Manassas, was only shifting, and not surmounting, a difficulty—that we would find the same enemy, and the same, or equal, intrenchments, at either place.

And when Lincoln appointed Grant lieutenant-general in 1864 he commented that "procrastination on the part of commanders" had always forced him to interfere with them in the past.[18]

Such was a habit of mind for Lincoln, but the analogy between enterprise in private life or economic life and enterprise in war was imperfect. Lincoln's outlook underestimated the value of maneuver and risked headlong attack of the sort that had undone General Burnside in December. Lincoln was stung by the defeat and therefore told Hooker to "avoid rashness," and in a memorandum of April 11 about the situation in Virginia he said, "I do not think we should take the disadvangtage of attacking him [the enemy] in his entrenchments." But, the commander-in-chief added, "If he weakens himself, pitch into him."[19]

Lincoln mentioned entrenchments but not terrain. His strategic ideas assumed that all other things were equal, but terrain is never equal. It varies in ways critical to combat from kilometer to kilometer. To realize the problem in command relations posed by an aggressive president in

Washington and a general miles away facing an enemy in Virginia, consider this episode from the time of the Chancellorsville campaign. General-in-Chief Henry W. Halleck had by the spring of 1863 completely appropriated the president's idea of the proper strategy. Halleck emphasized attack, he pointed to Lee's army rather than Richmond as the objective, and he was always imitating Lincoln's penchant for reminding generals that the external factors they invoked as excuses for failure or inaction somehow did not cause the Confederate generals to fail or be inactive. Thus Lincoln had scolded McClellan by asking him, "Are you not over-cautious when you assume that you can not do what the enemy is constantly doing? Should you not claim to be at least his equal in prowess, and act upon the claim? . . . It is all easy if our troops march as well as the enemy; and it is unmanly to say they can not do it." Imitating his boss the president, Halleck on April 24, 1863, scolded General B. S. Roberts, who had complained that heavy rain made roads impassable, "I do not understand how the roads there are impassable to you, when, by your own account, they are passable enough to the enemy. If you cannot drive the enemy out, we will seek someone who can." General Roberts replied with cool understatement: "The roads the enemy has passed are the mountain roads. Those I must move over are in the valley, and I have never seen any in so impassable a condition. I shall fail in nothing that is possible."[20] Terrain was all important and it could only be seen close-up. Maneuver cannot always be fairly compared to "procrastination," and local circumstances are not mere excuses for inaction; they can in fact affect opposing armies unequally, as the Wilderness did.

Hooker lost his battle in the Wilderness for obvious and now famous reasons, the main one being that he fought it in a wilderness where the Confederates' superior knowledge of the local area led Stonewall Jackson to a little-known road, hidden in the forest, that he could use to flank the right of the Union army. Operating where the people conveyed information to the Confederates but not to the Union, Jackson learned about the road through his chaplain Tucker Lacy, whose brother lived nearby. When Jackson objected that the route would take him too close to the Union pickets, Lacy assured him that a man named Charles Wellford, who lived at nearby Catherine Furnace, knew alternate routes. Jedediah Hotchkiss, Jackson's mapmaker, went to see Wellford, and the rest is history.[21] The Wilderness itself helped obscure Jackson's movements from Union observation. Jackson's army fell on the unsuspecting Yankees and put to flight Hooker's right wing, the Eleventh Corps, commanded by General O. O. Howard. The Wilderness gave Lee victory and guaranteed Union defeat.

But what of Lincoln's other virtue as commander-in-chief, his personal courage? It too could be a disadvantage, for it exacerbated a problem the roots of which lay too deep for mid-nineteenth-century American culture to deal with. It affected Union and Confederate alike, both the presidents of the belligerent regions, and probably all classes of soldiers but especially the officer class. It was a major problem of Joseph Hooker and without doubt diminished his performance at Chancellorsville. Courage became a vice when it led to "macho-ism," an exaggerated aggressiveness that was part of what might be called the cult of manliness.[22]

Nearly every male of the Victorian era seemed infected by this cult to some degree, and in fairness it must be said that Abraham Lincoln probably had a less severe case than average. He once advocated limited woman's suffrage. He showed little fondness for hunting large wild animals. He once protected a friendly Indian from murder at the hands of frontier militiamen. When challenged to a duel by the Democratic politician James Shields, Lincoln set such absurd conditions as to undercut the atmosphere of honor that was supposed to surround the affair. As a congressman, he had sharply criticized an American war. He once called military glory "that attractive rainbow, that rises in showers of blood—that serpent's eye, that charms to destroy."[23]

But a freak occurrence on the eve of Lincoln's presidency served to make him turn his back on his personal heritage of non-violence. Warned of an assassination plot in Baltimore on the route to his inauguration in 1861, the president-elect heeded the advice of others and sneaked into Washington in the dark of night and in disguise. It was surely a mistake. The contempt it aroused among southerners was later worth regiments in morale to the Confederacy—even the Yankee president looked like a coward.[24] It afforded a golden opportunity for journalistic abuse and cartoon scorn in the North, giving Lincoln an image of personal cowardliness that he had not lived down even by 1864, when he still appeared in at least one political cartoon dressed in tartan cap and cape (his alleged disguise for the night ride into Washington).[25] The tragic result was to cause Lincoln ever after to ignore pleas for personal caution in exposing himself to possible assassination. And when Jubal Early's Confederates reached the District of Columbia in the summer of 1864, the president felt compelled to expose himself needlessly to Confederate fire on the parapet of a fort at a spot known to Rebel sharpshooters. The soldiers in the area were appalled by Lincoln's foolish behavior.

Lincoln's personal political problem of manliness served to accelerate the natural tendency of the age to bring out the foolhardy worst in its men, especially its officer class. Soldiers must be courageous, officers and men alike, but unnecessary exposure to danger gained nothing. And as John Keegan has reminded us, there was less reason than ever in the American Civil War for generals to expose themselves to enemy fire. The greatly expanded scale of warfare demanded talents apparently more rarely found than courage: organizational skills and administrative abililities of a high order.[26] Hooker possessed such talents, incidentally, and he needed them. For example, the camp of the Army of the Potomac before it crossed the Rappahannock to Chancellorsville stretched out nearly one hundred miles in circumference. That meant that it was some thirty-two miles in diameter, and in practical terms, it took forty-eight hours—two days—to get a message from one end to the other.[27]

Keegan calls the generalship required for such enormous organizations a "new style of command," best exercised far to the rear of the battle lines, but in his description of it he leaves out one important factor in those who had to exemplify this new style: the feelings of the generals themselves. They had been trained to be courageous men, exemplarily so, and it embarrassed them to ask others to take risks that they did not share themselves. George McClellan expressed the agony best, in a letter to his wife after the Seven Days' battles at the end of July 1862: "I had no rest, no peace, except when in front with my men. The duties of my position are such as often to make it necessary for me to remain in the rear—it is an awful thing." Many other generals issued similar statements professing a desire to share the dangers with their men. Even Jefferson Davis felt this guilt powerfully. In a general order issued on February 10, 1864, the Confederate president proclaimed: "Soldiers! By your will (for you and the people are but one) I have been placed in a position which debars me from sharing your dangers, your sufferings, and your privations in the field."[28]

The behavioral result of such guilt is well known. Civil War officers threw themselves into a 15 percent greater casualty rate than enlisted men, and generals suffered the highest rate of all: they stood a 50 percent greater chance of becoming a casualty than did the ordinary soldiers.[29] Officers fell in such numbers because they exposed themselves recklessly to enemy fire on reconnaissance (like Stonewall Jackson at Chancellorsville), because they placed their headquarters too close to the action (like Hooker at Chancellorsville), and sometimes because they *wanted* to get hurt. They did such things because they were imbued with the cult of manliness in its exaggerated battlefield form. Military officers carried this aspect of Victorian culture to its counterproductive extreme.[30]

Although Hooker, who was not very articulate, did not express the guilt as McClellan did in a letter, he certainly evinced the behavior and the attitudes that often led to being shot in battle. And he was, of course. His horse was shot once at Yorktown. Hooker himself was shot at Antietam and taken from the field at a crucial moment. Once again, at Chancellorsville, Hooker was wounded, this time stunned when an artillery projectile struck the porch of the house that served as his headquarters and a piece of the porch in turn struck him. He had already lost confidence, but surely the wound did not *enhance* his judgment in the later parts of the battle.

Joseph Hooker was a war lover. Combat seems almost to have given him, to invoke modern slang, "a high." In his unfinished report of the battle of Antietam, Hooker said: "The whole morning had been one of unusual animation to me and fraught with the grandest events. The conduct of my troops was sublime, and the occasion almost lifted me to the skies, and its memoirs will ever remain near me." He recalled that the action ended for him when he was removed from his saddle "in the act of falling out of it from loss of blood, having been previously struck without my knowledge."[31] He had been shot in the foot and was either so excited he did not notice or so bent on showing his manliness that he pretended not to.

Hooker invited the wound at Chancellorsville by bravado. One journalist reported that the general stood calmly smoking a cigar on the porch of the house knowing full well he was in range of the Confederate artillery at Hazel Grove. When the projectile struck that caused his injury, Hooker was bending over the railing of the porch to listen to a report from an orderly.[32]

Although he once said he despised his nickname "Fighting Joe" because it made him sound like a highwayman, in fact Hooker's behavior fully lived up to the sobriquet. And so did his attitudes. His language was marked by braggadocio and semantic swagger. Some people at the time, including Lincoln, who was always studiously modest, were embarrassed by this, and almost everyone writing about Chancellorsville since has been embarrassed by Hooker's bragging. He also indulged an unusual language of contempt for the enemy. Writing to General George Stoneman on the eve of the campaign, Hooker, in orders drafted by an aide, urged, liked an overexcited coach at half-time, "If you cannot cut off from his column large slices, the General desires that you will not fail to take small ones. Let your watchword be fight, and let all your orders be fight, fight, fight, bearing in mind that time is as valuable to the General as the rebel carcasses." "Carcass" is the term used for the dead body of an animal, not a man, and such language, surprisingly, was rare in the dispatches of Civil War generals: Hooker had an advanced case of the cult of manliness.[33] In another note sent on the eve of the campaign to a general in another theater, Hooker said: "You must be patient with me. I must play with these devils before I can spring."[34] The twentieth century is famous for demonizing its enemies in wars, and this sounds mild to us, but the nineteenth century rarely did it for enemies of the same race and cultural stock. Such degrading language was reserved to enlisted men or for officers to use in reference to Indians.

After the war, in a rambling newspaper interview, Hooker blamed General O. O. Howard, commander of the Union right flank that collapsed under Jackson's assault, for the defeat at Chancellorsville. When the reporter asked whether Howard was a brave man, Hooker replied: "Howard is a very *queer* man . . . . He was always a woman among troops. If he was not born in petticoats, he ought to have been, and ought to wear them. He was always taken up with Sunday Schools and the temperanth cause. Those things are all very good, you know, but have nothing to do with commanding army corps."[35]

General Hooker's attitude could be said to have been infectious if it were not the case that most generals were already infected without his being the carrier. His chief of staff, General Daniel Butterfield, a lawyer rather than a professional soldier, toward the end of the battle of Chancellorsville, implored his commander to allow him to leave headquarters and join General John Sedgwick's forces for the remainder of the fight; he expressed himself as feeling "heartsick at not being permitted to be on the actual field, to share the fate and fortune of this army and my general."[36] By this time, of course, the fate of his general was to have been wounded. The desire to expose oneself recklessly when defeat threatened was extremely powerful and widespread and accounted for heaven knows how many officers' deaths. (We know only about the ones who lived to describe it afterward.) General Darius N. Couch, commander of the Second Corps in the Army of the Potomac, disagreed vehemently with Hooker's direction of the army, so much so that he requested and received a transfer after Chancellorsville. But the result of his frustration on the Virginia battlefield was thus described by Francis Amasa Walker, who admired Couch and thought little of Hooker's fighting abilities: "The orders he received were executed with energy and dispatch; and he even sought to find, in the reckless exposure of his own life, relief from the terrible sense that his own troops and the other gallant corps around him were being aimlessly sacrificed." Couch nearly got his relief: he was shot twice and his horse was killed by enemy fire.[37]

The most remarkable exemplar of this suicidal behavior at Chancellorsville was that general in petticoats, O. O. Howard. Howard had already lost an arm in battle in 1862, and, as his corps fled in uncontrollable panic from Jackson's assault on May 2, 1863, he recalled later: "I felt . . . that I wanted to die. It was the only time I ever weakened that way in my life, before or since, but that night I did all in my power to remedy the mistake, and I sought death everywhere I could find an excuse to go on the field."[38]

Neither Hooker nor Howard needed any encouragement in such attitudes, but it can be said that President Lincoln did little to discourage them. For example, recall Lincoln's language to McClellan quoted earlier: "It is *unmanly* to say they can not do it." In an age of strict separation of gender roles, the accusation of "unmanly" behavior constituted strong language—and a goad to recklessness.

Eventually the armies came out of the Wilderness, and then perseverance and pressure of attack prevailed for the Union cause. Before that happened, Joseph Hooker had been consigned to failure. There is no need to refurbish his reputation, for he did not perform well in the spring of 1863. But he ought to be given a fair shake. In Ulysses Grant's campaign a year later on that same ground—in a battle that goes by the name "Wilderness"—Grant lost more than double the number of Lee's casualties, losses that jeopardized President Lincoln's re-election in the fall. The Wilderness was Grant's excuse for losing. "No great battle ever took place before on such ground," said A. A. Humphreys in writing about the Wilderness battle for Scribners' Campaigns of the Civil War series in the 1880s.[39] But in fact

Chancellorsville was fought there too, and Humphreys was in Hooker's army fighting at the time!

Hooker is the victim of an unfortunate accident of naming battles. His most famous battle and defeat should not be called "Chancellorsville." There was no Chancellorsville—there was only one house at the site, standing in a clearing in the Wilderness.[40] Hooker's battle should be called what the battle was called a year later when Grant fought on the same ground, the battle of the Wilderness. Hooker's defeat would be First Wilderness, like First Bull Run, and thus the explanation of the general's failure would be built in to the battle's very name—and like Grant, Hooker might be forgiven his failure in impossible terrain.

Why was this not ever done? It was, briefly, at the end of the nineteenth century—by Augustus Choate Hamlin, whose influential book, *The Battle of Chancellorsville,* was published privately in 1896. He set out to vindicate the Eleventh Corps from charges that it fled from Jackson's attack because this Union corps had so many Germans in it. Hamlin cleverly fit the Union defeat and the arrogance and pride of the West Pointers who refused to fortify the flank, into the oldest American military myth, the myth of Braddock's defeat:

> . . . Jackson's collection of fighters, trudging along the woods and its by-paths, would certainly have presented a curious appearance to a martinet critic of any of the military schools of Europe. . . . But a closer and keener look, would have soon convinced him that outward appearances do not always indicate the true measure of the soldier, and he would soon have seen that this shabby-looking and apparently undis-

> ciplined rabble would, at a signal from their trusty leader, be transformed into a resolute army, more than a match for any equal number of the best troops of the European armies in the singular contest about to commence. And it may be affirmed that thirty thousand of these European troops would have been as helpless before them as Braddock and his British regulars were before the French and unseen Indians in the woods near Fort Duquesne, in colonial times.

He went on:

> The Wilderness, with its almost impenetrable thickets, was a great and natural fortress for the lightly armed and lightly clad Confederates. And the circumstances of the conflict recall the remarks made at the time of the Revolution of 1775, when it was said in England, that "The old system of tactics is out of place, nor could the capacity of the Americans for resistance be determined by any known rule of war. They will long shun an open field, every thicket will be an ambuscade of partisans, every stone wall a hiding place for sharp-shooters, every swamp a fortress, the boundless woods an impracticable barrier."
>
> And so it proved, for the rebel in his faded uniform was almost invisible in the woods, and his skill as a marksman, his knowledge in bushcraft, certainly compensated largely for a considerable inequality in numbers, and in the thickets of Chancellorsville, and later, in the Wilderness, the rebel soldier was certainly superior to his antagonist, man for man, courage reckoned an equal.[41]

With this interpretation available to explain Chancellorsville by invoking the greatest myth of all of American military history—that what this country did to gain military success was to jettison formal European rules and adopt the individualism, camouflage, and marksmanship of the Indian in the wilderness—why did it turn out that in fact wilderness was little emphasized in twentieth-century accounts of the battle and instead the emphasis was placed on Hooker's loss of initiative? There are at least three reasons: first, both sides were Americans, and it would be difficult to attribute a loss of woodcraft skills to the limited industrialization of the North in 1863; second, most Americans simply cannot bear to hear it said that they are no good at fighting in the woods; this argument stood on its head a myth so powerfully linked with the identity of America as Nature's nation that hardly anyone was going to embrace it lovingly; and, third, as Lincoln's military reputation rose, how could an interpretation that emphasized the role of wilderness in Union failure be squared with a commander-in-chief who was, as the subtitles of various biographies proclaimed, "The Backwoods Boy" or "The Railsplitter"?

The Wilderness interpretation of Chancellorsville died aborning, and it may not merit revival. But in losing our sense of the importance of terrain by rejecting this argument, we lost something essential to understanding the American Civil War. We also tilted the scales a little too much in Lincoln's favor as commander-in-chief. This son of the soil had long since left his pastoral roots, and in doing so he may have lost an appreciation for terrain as a factor in war.[42]

3

"Unfinished Work": Lincoln, Meade, and Gettysburg

Gabor S. Boritt

THE SNAPPING TURTLE AND THE COMMANDER-IN-CHIEF

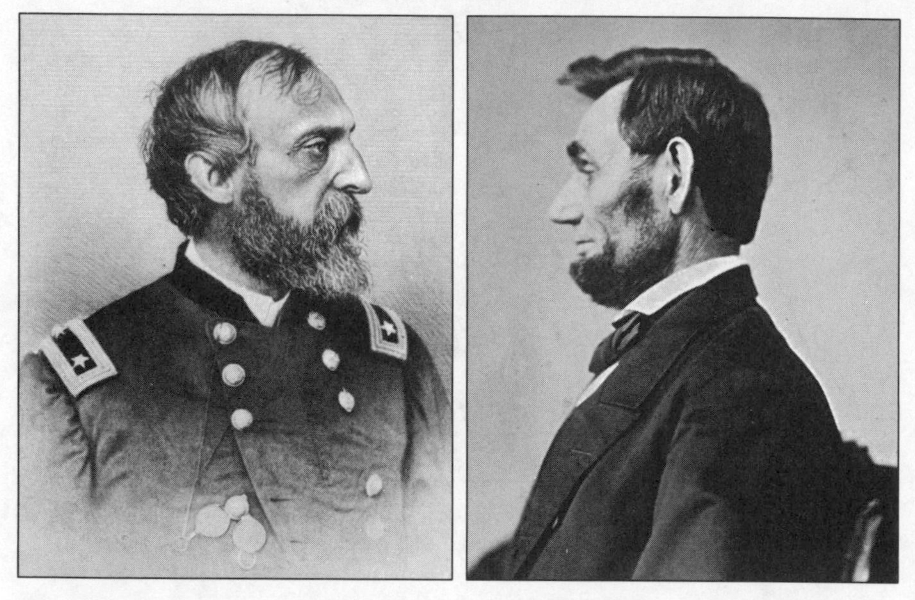

Steel engraving of General Meade by Charles B. Hall and photograph of Lincoln by Alexander Gardner, 1863. *(Engraving from Isaac R. Pennypacker,* Great Commanders: General Meade *(1901); Lincoln photograph courtesy of James Mellon)*

"COME TO WASHINGTON," Lincoln telegraphed his son, Robert, on July 11, 1863. The president's oldest child was not close to his father during the war years. He was at an awkward age when sons must distance themselves from their fathers as they become men. Lincoln called for Robert because his mother, Mary Todd, was gravely ill.[1] On July 2, she had had a carriage accident. The circumstances were not made clear, and historians still don't know whether Mrs. Lincoln had an authentic accident or was the victim of sabotage—with the bolts of the carriage seat intentionally loosened—sabotage intended for the commander-in-chief. In either case, the battle of Gettysburg in all probability saved Lincoln from injury. Instead of taking a ride with his wife, he spent much of that day at the telegraph office of the War Department, anxiously waiting for news from Pennsylvania.

Then the good word came from Gettysburg, and Lincoln issued a public statement:

> The President announces to the country that news from the Army of the Potomac, up to 10 P.M. of the 3rd is such as to cover that Army with the highest honor, to promise a great success to the cause of the Union, and to claim the condolence of all for the many gallant fallen. And that for this, he especially desires that on this day, He whose will, not ours, should ever be done, be everywhere remembered and reverenced with profoundest gratitude.[2]

A day earlier, Lincoln had telegraphed his son at Harvard University, knowing that Robert might see in the newspapers a mention of Mrs. Lincoln's accident. "Don't be uneasy. Your mother very slightly hurt by her fall." Knowing also that presidential wishes were thwarted at all levels, from telegraph clerks to generals, he added: "Please send at once." Then, her wound having become infected, Mrs. Lincoln took a turn for the worse. She seemed in danger. And so during the next ten days, Lincoln went through hell, deeply anguished about the fate of the two armies in southern Pennsylvania, believing that the outcome of the Civil War itself hung in the balance, and also deeply anguished over his wife. No wonder he sent the terse telegram to Robert on the 11th: "Come to Washington." Another telegram survives from the 14th: "Why do I hear no more of you?" That's all. What communications in between are missing we do not know, but Robert had been on his way and arrived at the White House on the same day.[3]

The meeting with his father that followed, he would

never forget. Mrs. Lincoln, in the meantime, started to get better. But at high noon on the 14th, word came from General Meade that Lee's army had crossed the Potomac into Virginia. The President felt devastated. Robert must have arrived soon after, for he found his father "in tears," the only time in his life that he had seen "his father cry." Asked what was the matter, Lincoln sighed that the "chance of ending this bitter struggle is lost." Lee had escaped. The father also shared confidential details with the son. By the evening of the next day, Robert could not help but relate some of it to John Hay, his closest chum in Washington, how the president avowed that "If I had gone up there I could have whipped them myself." And Hay, who as private secretary had spent the past ten anxious days with "the tycoon," as he sometimes referred to his boss, noted triumphantly in his diary: "I know he had that idea." Lincoln had the urge to take to the field himself after Gettysburg, though he would blurt this out only to his son.[4]

If Abraham Lincoln cried at the end of what was called the Maryland campaign, George Gordon Meade had a somewhat less emotional response, though his nickname was "the old snapping turtle." His son George, a captain on his father's staff, had written home to his mother on July 7 that "Papa will end the war." In part, young Meade bespoke his admiration for his father; in part, he took his clue from the general himself, who on July 9 himself informed Washington: "I think the decisive battle of the war will be fought in a few days." When to Meade's surprise—though not to Lincoln's—Lee crossed over to Virginia without a fight, the commander of the Army of the Potomac complained about his superiors in a letter to his wife, Margaret: "they . . .

insist on my continuing to try to do what I know in advance it is impossible to do."[5] The Gettysburg campaign was over.

Lincoln with his tears, Meade with his lament, drew the battle lines for historians that stand to this day. This essay is only the latest, respectful attempt to join the fray.

Earlier, when Lee invaded Pennsylvania, Lincoln was elated. His worried wife, visiting Philadelphia at the time, must have wondered whether she should rush home to Washington. The newspapers cried great alarm, after all, but Lincoln wrote her calmly: "I do not think the raid into Pennsylvania amounts to anything at all." In reassuring his wife, he did not go on to explain that he actually placed potentially great importance on Lee's movement, for, as he said elsewhere, it presented "the best opportunity we have had since the war began." His well-considered judgment was the fruit of having so ably learned military science on the job as commander-in-chief. By the time he felt strong enough in his learning to fire McClellan for the second time, in the fall of 1862, he could explain matters clearly to his general with the "slows." "You dread his [Lee's] going into Pennsylvania. But if he does so in full force, he gives up his communications . . . and you have nothing to do but to follow, and ruin him. . . . If we can not beat him when he bears the wastage of coming to us, we never can. . . . This proposition is simple truth. . . . We should not so operate as to merely drive him away . . . we must beat him somewhere, or fail finally . . . ."[6]

If by September of 1862 Lincoln could propose permitting the enemy to go to Harrisburg and Philadelphia, and so "end the war" (except that he worried about demoralization and mutiny in the Keystone state), it is not surprising that after Lee's invasion began in 1863, people found him in "excellent spirits." The contrast between the northern public's near panic, the headlines about the Rebel Invasion on the one hand, and the president's elation on the other is striking. As he explained many weeks later, "I had always believed . . . that the main rebel army going North of the Potomac, could never return, if well attended to . . . ." "We cannot help beating them," he exulted to the cabinet on June 26, but he was worried that the commander of the army, Joseph Hooker, might turn out to be another McClellan. "How much depends in military matters on one master mind," he mused philosophically, as if statesmanship was altogether different.[7]

After the disaster at Chancellorsville, many in the Army of the Potomac hoped that George Gordon Meade might be appointed in Hooker's place. For the moment the president refused to make a change though he kept "studying Meade." Then however with Hooker looking more and more like McClellan the closer he got to Lee, Lincoln and Stanton decided to replace him with the Pennsylvanian. He "will fight well on his own dunghill," the president was remembered to have said. It was not a highly enthusiastic endorsement.[8]

When a year later Lincoln ran for re-election, many thought his best argument in his own favor—in favor of a president who had failed to win the war—was the one about changing horses in midstream, how that was a bad idea. Of course, the same went for changing the commander of an army at the critical moment of a campaign. As Lincoln contemplated what to do, not surprisingly the favored proverb came to his tongue; and also to those of others. But he judged that worse than swapping horses was a demoralized Hooker, acting perhaps irresponsibly. So the president made his move.

Colonel James A. Hardie was dispatched from Washington to find Meade, which he did at 3:00 a.m. on the 28th of June. The scene is famous: The Colonel waking the commander of the Fifth Corps and telling him that he brought "trouble." Meade, barely awake, assumed himself to be relieved or even under arrest by order of General Hooker. Instead, Meade was the new commander of the army.[9]

The general's war record was rock-solid, from the Peninsula—where he suffered wounds—through Antietam, Fredericksburg, and Chancellorsville, where he had been among the angry stalwarts who wanted to stay and fight, only to be overruled by Hooker and the majority. "I am tired of this playing war *without risks,*" he had written to his wife at the start of 1863. Earlier he had condemned McClellan to her for losing "the greatest chance any man ever had on this continent." He went on to say in 1863 that McClellan was "always waiting" to have every thing just as he wanted before he would attack." He lost great opportunities "for *want* of *nerve* to run what he considered risks. Such a general will never command success, though he may avoid disaster." After Chancellorsville, where "Fighting Joe" Hooker, though outnumbering Lee more than two to one, met disaster, another insight seems to have come to Meade. Again writing to Margaret, he lamented that Hooker failed to "show his fighting qualities at the pinch—He was more cautious . . . even than McClellan—thus proving that a man may talk very big, when he has no responsibility—but that it is quite a different thing, acting when you are responsible . . . ." Not surprisingly Meade wrote this to his wife just after he, and other corps commanders, had lunch with Lincoln.[10]

How much of this sank in for Meade? Did any of these words come back to haunt him after Gettysburg? For strong, clear statements and brave deeds notwithstanding, once in command of the Army of the Potomac, Meade proved to be "a *juste miliau* [sic] man"—a middle of the roader—as he described himself many years earlier.[11] He was not that in all matters. After all he would institute Friday executions for each of his army corps, and he could in battle strike a cowering soldier with the flat of his sabre. He was the "old snapping turtle." Yet at the core, he showed himself to be unimaginative, cautious, conservatively suspicious of the masses—including the volunteer soldiers—highly professional, and at his best defending his own ground or (to quote again words stuck on him via a very different kind of a man) "his own dunghill."

At Gettysburg, Meade won a brilliant victory. To use the cliché, the man and the hour had met. The terrain, Lee's need to attack, Meade's solid, cautious temperament and, of course, as always in battle, luck combined to help the soldiers and officers produce the northern triumph.

In Washington, Lincoln issued his grateful proclamation noting that the army covered itself with "the highest honor" at Gettysburg. But to the careful reader, what jumps out of the document is the "promise," the immediate future expectation of "a great success." This implied that, Gettysburg notwithstanding, a great success there had not yet been. The president wanted the capture or destruction of the largest Confederate army. Even when a crowd came to serenade him at the White House to celebrate both Gettysburg and the surrender of Vicksburg (with its 30,000-man garrison), and Lincoln gave an awkwardly exultant impromptu response, he could not help saying: "These are trying occasions, not only [daily?] in success, but for the want of success."[12] The Army of Northern Virginia was still in the field, in spite of Gettysburg, and the Confederate cause very much alive. Thus once again, the reactions of the president and the people diverged sharply. Before Gettysburg, Lincoln rejoiced while people wailed fearfully; afterwards, shouts of "victory" resounded over the North while the president refused to use the word and looked for the victory yet to come.

In private, Meade exulted. Why should he not have?—the Army of the Potomac had at last defeated the Army of Northern Virginia. To his wife he wrote: "It was a grand battle, and is in my judgment a most decided victory, though I did not annihilate or *bag* the Confederate Army. . . . The [Union] army are in the hightest spirits, and of course I am a great man."[13] And he, too, issued a public proclamation, General Order No. 68:

> The commanding general, in behalf of the country, thanks the Army of the Potomac for the glorious result of the recent operations. An enemy, superior in numbers, and flushed with the pride of a successful invasion, attempted to overcome and destroy this army. Utterly baffled and defeated, he has now withdrawn from the contest. The privations and fatigue the army has endured, and the heroic courage and gallantry it has displayed, will be matters of history, to be ever remembered.

So much was typical army language, and surely the soldiers and much of the North welcomed it. So it was with

Meade's next, brief paragraph, though to a discerning few it also revealed a specific cast of mind about military objectives. "Our task is not yet accomplished," Meade explained; he wanted more of his victorious troops: "to drive from our soil every vestige of the presence of the invader." [14]

In Washington, the president continued to hover at the telegraph office. He was there when Meade's Order No. 68 came in. He reached for it eagerly. As later recalled, he showed no special emotion reading his general's exaggerated claim of having defeated "superior" numbers. But when Lincoln came to driving the invaders "from our soil," his hands fell on his knees, and his face and voice bespoke anguish: "Drive the invaders from our soil! My God! Is that all?" Ten days later in the presence of his diary-keeping secretary, John Hay, Lincoln was equally bitter: "This is a dreadful [notion] . . . . Will our generals never get that idea out of their heads? The whole country is our soil." [15]

Lincoln would not fight over words unless they had practical consequences. This time he judged that they did: he wanted Lee's army captured or largely destroyed. What parts of Pennsylvania or Virginia were held by whichever side paled to insignificance by comparison. Like so many Americans, the president believed in the one great Clausewitzian battle (though he never read the Prussian theorist) and hoped it would bring the terrible war to an end.

Meade did not disagree with Clausewitz or Lincoln on this, but seemed well satisfied by his achievement at Gettysburg and was in no hurry to endanger either Washington or his own newly won reputation by offering a battle that could be lost. He was all too aware of fickle fortune. Lincoln, in turn, was too unaware of the strength the Army of

Northern Virginia still retained even in defeat, and seemed to dismiss entirely the difficulties faced by the Army of the Potomac.

Confederate Porter Alexander agreed with Lincoln on some things. "As it was the Fourth of July, there was an idea that Meade would be inspired to try and win a real victory," he wrote in his memoirs.[16] Meade may well have been quite right in not attacking, but he did not think aggressively in any case. (He failed even to consider seizing the Fairfield gap, a few miles west of Gettysburg, in an attempt to prevent the Confederates from withdrawing that way.) The task may have been overwhelming, best not thought of at all. Still, others acted with more vigor. Even ordinary Pennsylvanians, for example, were bold enough to attack Lee's retreating wagon train of misery and knock out the spokes of wagon wheels. Or there was General William H. French who, without orders from Meade, sent a detachment to destroy Lee's sole pontoon bridge over the Potomac, at Falling Waters. Whether deep down Meade truly wished to face Lee again on a battlefield north of the river, the opportunity would be there.

In the dark of the stormy night of the 4th, Lee's army left Gettysburg, assured several hours of head start over the pursuers. In fact, Meade would not manage to make a real start until July 7. By then, Longstreet's corps, leading the Confederates, had arrived at Williamsport, the seventeen-mile-long wagon train of the wounded and the ten-mile-long baggage train having preceded it. The Army of Northern Virginia executed a potentially very dangerous retreat as a model maneuver. But then, with pontoons gone and the Potomac swollen, it had to stop. The southern soldiers appeared to have retained much of their fighting spirit, their tremendous losses notwithstanding. They may have been ready, even eager for a rematch. It is also reasonably clear, however, that the morale of their northern opponents had never been higher. For at Gettysburg they had lost not only many men but a good part of their awe of Lee's army. And Meade's army was getting help. Wrote Lt. Col. James H. Campbell of the 39th Pennsylvania Militia Infantry to his wife Juliet: "My dear wife—I am very well—it rains—Let it pour. I thank Heaven for it. It will swell the Potomac and the mountain streams. What will become of Lee's routed columns? Let it rain!"[17]

Lincoln, too, believed that the Army of Northern Virginia had been routed. And Meade seemed to him a better general than any of his predecessors. Yet was he going to be equal to the task before him? When July 5 went by, without word of renewed battle, Lincoln's doubts increased. With his youngest son Tad, he went to visit General Daniel E. Sickles, who had come to Washington to recuperate from the loss of a leg. The spunky congressman-turned-soldier gave the president an earful about Meade's deficiencies. Lincoln, however, understood politicians and no doubt discounted a fair part of what he heard. But July 5 also brought a prediction from the once Gettysburg-based college professor, and now general, Herman Haupt, that Meade would allow Lee to escape. The next day brought Haupt himself. Though some of the general's information was poor, his central point turned out to be prophetic. And Haupt had Lincoln's respect.

Actually, July 5 and 6 saw minor but potentially important action when Sedgwick's corps touched the rearguard of Lee's army and might have brought on battle close to Fairfield. But Sedgwick held back. At Williamsport and Hagerstown, the cavalry of Buford and Kilpatrick attacked the trains of wounded and supplies, while threatening the screen protecting Lee's approaching main columns. Imboden, the Confederate commander, later wrote that Lee's army could have been "ruined" here, but he exaggerated.[18] In any case, lack of Union coordination and solid Confederate defense saved the day for the graycoats.

If the Army of Northern Virginia showed superb qualities in executing a potentially dangerous retreat, it was aided by the total lack of aggressive pursuit except by cavalry. Meanwhile, as the North celebrated the great victory of Gettysburg, in Washington Lincoln showed the first signs of losing his temper. As various bits and pieces of information came in, the commander-in-chief wrote to Halleck: "These things all appear to me to be connected with a purpose to cover Baltimore and Washington, and to get the enemy across the river again without a further collision, and they do not appear connected with a purpose to prevent his crossing and to destroy him."[19]

On July 7, at last the Army of the Potomac showed what it could do as it moved rapidly after the Rebels, albeit not via the roads taken by them. Most units covered fifteen to twenty miles that day, but some as many as thirty miles. The feat was all the more impressive because the rains came, at times in torrents. Meade could make his army move fast, though—and the unkind thought cannot be escaped—not only the good roads but the enemy at a goodly distance helped.

In Washington, a gloomy Lincoln spoke to the Cabinet

his fear that "the old idea of driving the Rebels out of Pennsylvania and Maryland" yet remained. He was "anxious" to praise Meade and the army, but they needed an infusion of "the right tone and spirit."[20] Halleck, caught between an aggressive president and a cautious general, one day leaned this way, the next day that. And Secretary of Navy Gideon Welles, at least, thought the president failed to push Meade hard enough.

When on July 7 word arrived of Vicksburg's surrender, Lincoln's first thought was to send the news to Meade. Elated, the president used Halleck as the intermediary to ask that the general "complete his work" so ably advanced "thus far." The very words would reappear in the Gettysburg Address. Lincoln wanted the "literal or substantial destruction of Lee's army." If done, "the rebellion will be over." To make matters now fully clear, Meade received his appointment as brigadier in the regular army, dated to July 3—assuring high rank and pay even for peacetime. In this, historian Kenneth Williams saw unmistakable promise of another permanent promotion, to the highest rank in the army, if Meade would finish his "work." Porter Alexander thought it would have made Meade president of the United States. Nor were these the mere ruminations of latter-day pundits. Cincinnati newspaperman Whitelaw Reid reported a band serenading Meade on the evening of July 3 with the "significant melody—'Hail to the Chief.' " When someone remarked that the general was "in very great danger of being President of the United States," another officer advised: "Finish well this work well begun, and the position you have is better and prouder than President."[21] No shred of evidence exists that the possibility of a rival being created by Gettysburg ever troubled Lincoln. Nor is there any shred of evidence that the possibility enticed Meade. The strength, and weakness, of "*juste milieu*" men is that they are not so very ambitious. They know, and fear, the jealousy of the gods.

A crowd of serenaders came to the White House, too, on the evening of July 7. Lincoln spoke to them thoughts that in refined form would reach immortality when delivered at Gettysburg. His impromptu speech also showed the partly necessary gulf between his understanding of the battle and that of the general, who spoke of liberating Pennsylvania, "our soil."

"How long ago is it?" the president asked his audience, "eighty odd years—since on the Fourth of July for the first time in the history of the world a nation by its representatives, assembled and declared as a self-evident truth that 'all men are created equal.' That was the birthday of the United States of America." At times, Lincoln could be a terrible extemporaneous speaker. But the serenaders cheered enthusiastically all the same. So he went on to laud the victory at Vicksburg on July 4 and then turned to Gettysburg, showing he well understood what happened there: "a succession of battles . . . through three days, so rapidly fought that they might be called one great battle on the 1st, 2d, 3d of the month of July; and on the 4th the cohorts of those who opposed the declaration that 'all men are created equal,' turned 'tail and run.' " This time the cheers were long and continuous. Lincoln went on to evaluate his own speech frankly: "this is a glorious theme, and the occasion for a speech, but I am not prepared to make one worthy of the occasion." While celebrating with the happy serenaders the president also blurted out that this was not only a season of "success," but also a " 'trying' time for the want of success." As later, in the Gettysburg Address on November 19, Lincoln could not ignore that Meade's work, and so the nation's, remained unfinished.[22]

On July 8, the Army of the Potomac started to cross the South Mountains, but it took close to four days to face up against Lee's army only a few miles away. The Yankees probed, sparked skirmishes, and crept forward. Not until the morning of the 14th did Meade order a reconnaissance in force that could have become an actual attack. Alexander, the southern army's artillery chief, later wrote: "The enemy had pursued us as a mule goes on the chase of a grizzly bear—as if catching up with us was the last thing he wanted to do."[23] Actually, the mule never did quite catch up with the bear.

Alexander had ample opportunity to observe. The Army of Northern Virginia had been stopped by the loss of its pontoon bridge and the swollen Potomac on the 7th of July. The rains continued. Lee was a cornered and crippled tiger who had lost a very much larger proportion of his army than had his adversary, but a tiger still. He had to use all his genius, all his inner powers, and also the skills of the finest engineers he could muster to block the possible destruction of his army. Only one or two rafts could go back and forth over the Potomac. Supplies, wounded, and prisoners had to be ferried. A pontoon had to be built. Not until July 10 did he have sufficient artillery and presumably rifle ammunition. Lee's men had to build a definite line of defense. However ably this job was performed—often offering cross fire in front of marshy grounds—however stronger the fortifications grew as the days slipped by, however emphatically and to a degree self-servingly some Union officers, Meade included, would later praise its strength, Alexander (who would have welcomed a federal attack) remembered that "no very well defined and naturally strong" defensive line could be found. "We had to pick and choose, and string together."[24]

Yet through it all Lee stood as a mountain of strength. It is true that at times he seemed worried, even upset, and at moments some received the impression that the situation of the army was "precarious in the extreme." To Alexander he seemed more distorted than ever before or after. To his wife Lee wrote that "Had the river not unexectedly risen all would have been well with us. But God, in His all-wise providence, ruled otherwise . . . ." He reported to Jefferson Davis in the same manner but excluded the religious reference.[25] Mostly he showed confidence, specially as the days went by, even an eagerness to be attacked. In short, with the swollen Potomac behind him and no chance to retreat, he got ready for whatever might come. He was General Lee.

Meade knew that all too well. Mule and grizzly bear. One should sympathize with him. New to command, sustaining heavy losses at Gettysburg to a structurally weakened army, the general carried his burden the best he knew how. Fatigue enveloped him. Eleven days after taking command he wrote to his wife: "I have not changed my clothes [once], have not had a regular nights rest and many nights not a wink of sleep, and for several days did not even wash my face & hands—no regular food, and all the time, in a great state of mental anxiety—Indeed, I think I have lived as much in this time as in the last 30 years."[26] If he overestimated Confederate strength, he did so on the basis of the best intelligence. Two of his ablest comrades were gone: Winfield Hancock badly wounded, John Reynolds dead. From a partly misinformed Washington, Lincoln and Halleck put steady, even angry, pressure on Meade. His having to spar with his superiors, and deflect their partly unrealistic expectations took its toll, too. The specter of his fallen predecessors could never have been far from his mind. Reinforcements sent to him arrived but slowly. On July 8 Lincoln telegraphed Lorenzo Thomas, the Adjutant General then in Harrisburg where militia was being organized: "The forces you speak of, will be of no immagineable service, if they can not go forward with a little more expedition. . . . in my unprofessional opinion [they will] be quite as likely to capture the Man-in the Moon, as any part of Lee's Army."[27] At least when the commander-in-chief exploded toward others, Meade was being spared. In any case, the head of the Army of the Potomac (like so many officers on both sides) had all too low an opinion of the emergency troops, even when they fought close to home. He would not incorporate them even temporarily into his army.

Once Meade crossed the South Mountains and found Lee with his back to the still-cresting Potomac, he set about to prepare systematically for "the decisive battle of the war." The momentousness of this present had clearly rubbed off on him from Lincoln. The general worked very hard. He solved supply problems, lost veterans, gained reinforcements, reorganized, and ever so slowly crept toward the enemy. It would be better to fight here than on enemy territory, he had written to his wife on July 8. Good common sense. Two days earlier, one of his soldiers had written the same: "I hope we shall be able to destroy Lee and his Army before they get back across the Potomac for I do not want to chase them down through the state of Virginia again." Nor did Lincoln, nor most of the Army of the Potomac, which, Meade noted, was "in fine spirits." The majority of the men, another soldier wrote, considered Lee's "annihilation or capture certain."[28]

Lincoln, though, was far from sure. Since he believed that an immediate end of the war had now become possible, the stakes appeared to him all the greater. He rode a rollercoaster of emotions. A cipher clerk at the War Department's telegraph office remembered him walking "up and down the floor, his face grave and anxious, wringing his hands and showing every sign of distress. As the telegrams would come in he traced the positions of the two armies on the map. . . ." In public he remained calm and struck a cautious note, not wishing to raise expectations too high. To the Republican gubernatorial candidate in California, he wrote, "no doubt that Gen. Meade . . . beat Lee," but added that the latter was crossing the Potomac "closely pressed by Meade." Lincoln began to sense that if the Confederates escaped, to buoy northern morale it would be all the more important to focus on Gettysburg, and not its aftermath.[29] In private, however, he made desperate efforts to prod Meade into aggressive action before Lee escaped. Accordingly, perhaps on July 10, the president sent an extraordinary confidential order that only a crafty civilian could have devised. The text no longer appears to survive, but Robert Todd Lincoln reconstructed the substance of the words as early as 1872 for General Rush C. Hawkins, of Hawkins' Zouaves fame, by then American consul in Hamburg. Hawkins in turn committed all to paper on the spot:

"To Major General Meade, Commanding the Army of

the Potomac, You will follow up and attack General Lee as soon as possible before he can cross the river. If you fail, this dispatch will clear you from all responsibility and if you succeed you may destroy it."[30] Surely the president would have understood the insulting nature of such an order and intended with it in part to goad his conservative commander into action, and in part to absolve him from blame should an attack on Lee fail.

Lincoln spoke of the order to his son on July 14, 1863, the day the young man arrived in Washington to visit his injured mother. It was then that he saw his father with tears on his face. Presumably, when the president later wrote of not having "controlled events" but confessed "plainly that events have controlled me," in the back of his mind he carried the burden of a plethora of specific matters in addition to the large philosophical and theological considerations.[31] If young Robert could not but immediately blabber about the emotion-laden episode, while carousing with Hay, and spoke of his father's grief, he kept to himself longer about the tears and secret order. Perhaps not until the beginning of the next decade, when the still-young Robert was traveling in Europe, did he open up more fully. Certainly, the first written evidence seems to appear then. He never changed the essential parts of his account.[32]

Twenty years after Gettysburg, when Secretary of War, Robert wrote to an old family acquaintance and then a Lincoln biographer: "My recollection of the incident is perfectly distinct." He also added that he was either not told or did not remember "whether the communication from my father to General Meade was by letter, sent by the messenger, or by the telegraphic message."[33]

The probable courier of 1863 was no less a person than

the Vice President of the United States, Hannibal Hamlin, whose son Charles served with the Army. Lincoln had decided against going to the front himself, however tempted, but must have hoped that his second-in-command and a strong, provocative message would prove to be compelling. If Hamlin indeed carried the message, he turned out to be ineffectual, surely helping his removal from his post the following year. Meade's son wrote home only with contempt and incomprehension about the august visitor on July 11, 1863: "Vice President Hamlin has been here and went away this morning, what they sent him for, God only knows, he does not look as if he had an idea in his old beastly head." Some things General Meade shared with no one.[34]

In later years, having no independent proof, Robert Lincoln would not mention the order in public, much less permit its inclusion in a biography—however vivid and disturbing his own memory was. After all, in the post-war era, the issue was powerfully charged. Many people assumed that had Meade attacked Lee at Williamsport, the war would have come to an end more than a year and a half earlier than it actually did, saving hundreds of thousands of lives. Untold numbers of bereaved mothers, widows, and orphans of the latter part of 1863, 1864, and 1865 were then still alive. General Meade's son never heard of the order, and President Lincoln's son did not wish to engage in public controversy. Nor did John G. Nicolay and John Hay, the President's wartime secretaries and most eminent biographers in the nineteenth century. Nicolay could not find a copy of the order, and he may have been loath to believe that he had been excluded in 1863. Historians ever since have followed Nicolay and Hay's lead.

During his tenure at the War Department, Robert had the pleasure of happening upon James G. Fry, Civil War era provost marshal general, who was poking about for something like the lost order, or so Lincoln thought. Here seemed to be something of an independent corroboration. He also exchanged ideas with Meade critics Abner Doubleday and Herman Haupt, looking for proof. But only his own sharp recollection served as solid proof, and as the younger Lincoln grew old, he continued to refuse to go public. The ancient incident of his father's tears, however, gnawed at him all the same, and in private he continued to tell of it to a select few. He also gave the impression that someday he wanted the world to know, though "he did not precisely say so."[35] Indeed, in 1925, the year before he died, Robert Todd Lincoln again mentioned the matter. By then, the rumors surrounding the lost order had led to what must have been a forgery of the order being sold at auction. Historians may never find incontrovertible proof of Lincoln's order, but it is likely that the desperate president, believing more to be at stake than actually was, would have tried almost any approach.

Lincoln's secret order had no more effect than it would when he would try it again in the fall on Meade, albeit in a less direct manner. Meanwhile, during the hot, wet days of mid-July 1863, the president watched every day "with agonizing impatience." Hay reported, "hope struggling with fear." Meade was slow, steady, methodical, and Lincoln's emotions went up and down. On July 11, with the Army of the Potomac facing the Army of Northern Virginia, he basked "in the prospect of a brilliant success." Indeed, Lee's engineers still had much fortifying to do, and were just starting to send their Union prisoners to permanent captivity in the South. On the 12th, Meade wired his intention to attack on the morning as Lee continued his fortification. Reflecting the tycoon's view, Hay noted the next day that "nothing can save them, if Meade does his duty." But he added: "I doubt him. He is an engineer."[36] This was no humorous echo of Lincoln's comment about McClellan being a stationary engine, but rather a sober judgment. Indeed, the paucity of precise information weighed heavily on Meade's fine, engineering mind. And perhaps on the minds of others, too.

For Meade's telegram heralding the impending attack also contained an ominous phrase: "unless something intervenes."[37] The something was a council of war the commanding general called on the night of the 12th. He mildly proposed careful action; the majority of the corps commanders opposed it. Though historians will never fully know what happened, many of the generals appear to have deluded themselves that Lee would attack if they did not. Certainly the Army of the Potomac heavily fortified its own defensive line. It is also possible that a supposed Confederate "deserter" (an oft-used ruse), who earlier told tales of his army's refusal to withdraw over a completed bridge and also noted the men's eagerness to fight, still carried some influence. Certainly the Confederate position was strong, and risking defeat might risk the fruits of the battle already won.

So caution prevailed. When word reached the White House, Halleck shot back a harsh reply. He knew the president's mind. "You are strong enough to attack and defeat the enemy before he can effect a crossing. Act upon your own judgment and make your generals execute your order.

Call no council of war. It is proverbial that councils of war never fight. . . . Do not let the enemy escape."[38] By then Meade had ordered reconnoitering and, that done, substantially the same attack for the next day, the morning of July 14: reconnaissance in force to be converted to attack if practicable. Lee reinforced further, and since his pontoon was finally completed and the Potomac starting to subside, also readied for crossing the river to Virginia. Looking at the passive entrenching Yankees, he passed scorching judgment on them: "They have but little courage!"[39]

The morning of July 14 found Lincoln, again depressed, fearing that Meade, orders notwithstanding, "would do nothing."[40] One suspects Lincoln's intuition did not fail him—Meade might have found a way to stop short of a full attack—but this, too, we will never know. For when the Army of the Potomac moved, it found the Confederate defenses empty.

The river crossing was one of the tensest moments of Lee's military career. But he planned and performed unimpeachably. One corps, Ewell's, was to cross the ford at Williamsport, with much of Stuart's cavalry bringing up the rear. The other two corps, Longstreet's and A. P. Hill's, were to use the pontoon bridge at Falling Waters. Nature intervened once more. The skies opened again, providing camouflage but also making the Confederate crossing as miserable an experience as the retreat from Gettysburg. Once again in stealth, under cover of the night, the Army of Northern Virginia struggled to move. Surely Longstreet exaggerated when he remembered that "the best standing points were ankle deep in mud, and the roads half-way to the knee . . . ." Slowly, the army attempted to crawl to safety.[41]

Dawn found Longstreet and Hill still on the north side of the river. Some Union horsemen noticed as early as 3:00 a.m. that the Confederates were moving, and the Sixth Corps began the pursuit before Meade's appointed time of 7:00 a.m. Too little, too late. Meade did not order a general pursuit until 8:30.

No adequate explanation has ever been offered as to why he designated such a late hour for the beginning of action. The tardy start turned out to be of importance, for earlier substantial losses might have been inflicted on Longstreet and Hill. But the late start seemed to fit, to the bitter end, the spirit—or lack thereof—of the generals' pursuit of Lee. The common soldiers may have been ready enough to act vigorously, however bloody a field awaited them; so were some of their commanders. But not enough of them were. For long hours on the night of July 13 and the morning of the 14th, the Army of Northern Virginia had dangerously exposed itself as it streaked toward safety. But the Army of the Potomac slept, its commander confident that his well-entrenched opponents would stay, and give battle.

With nerves worn raw, Lee supervised his army's retreat. At last the thunder of a heavy gun announced his opponents were awake. "There!" he exclaimed, perhaps with relief, "I was expecting it—the beginning of the attack!" By then, only his rear division was still exposed. It took some casualties, and as the finest historian of the campaign wrote: "Ironically, the Confederate Henry Heth, who had opened hostilities at Gettysburg, also fired the closing volleys fourteen days later at Falling Waters—and in both instances he was caught unprepared."[42]

In Washington, Lincoln was prepared. Yet not even a "great man" can fully prepare for the magnitude of disappointment that now faced him. When the news came, a Cabinet meeting was about to get under way. The president cancelled it immediately. No one was "in a right frame of mind," he thought. "He was not." As members walked away, he hurried to catch up with Gideon Welles. Lincoln had to unburden himself of a little of the pain. His "voice and countenance," the Navy Secretary wrote in his diary, "I shall never forget."[43]

Lincoln said that "he had dreaded yet expected this." Or as he later put it to John Hay, "nothing I could say or do could make the Army move." "We had them within our grasp. We had only to stretch forth our hands & they were ours." In his moment of almost total despair with Welles, Lincoln even succumbed to the great American disease of seeing conspiracy. "There is bad faith somewhere. . . . What does it mean, Mr. Welles? Great God, what does it mean?"[44] The president may have spoken aloud his ugliest thoughts, knowing they would fall on fertile ears. For the navy secretary saw clearly "a great plan" that involved Lee's invasion; Confederate Vice President Alexander H. Stephens's simultaneous attempt to come to Washington to negotiate—rejected by Lincoln and advisers; and the riots that had just broken out in New York and elsewhere. Welles wondered whether "our own officers" were part of the malevolent scheme. In short, he saw not only "weakness" but "wickedness." At least Lincoln, even in his bottomless misery, did not dare go quite so far as to ask explicitly whether the Army of the Potomac had traitors in high places. On the same day his son Robert arrived at the White House.

The tears came, and the words. That, too, helped. Shortly after, Welles saw the president again, lying on a sofa in the War Department, a "subdued and sad" man, but "calm and resolute."[45]

Meade of course, was a patriot, not a traitor. But neither in his politics nor matters military could he fully share Lincoln's views, which pointed toward relentless, unyielding war. Meade moved with the times, did so with care, but may have been temperamentally unfit to do what a Grant or a Sherman would to end the war. It did not help Meade to have his wife's sister married to Confederate general Henry A. Wise, a former governor of Virginia; have a brother-in-law die in the Confederate Navy; and have the children of another sister, the mistress of a Mississippi plantation, lose her sons to the war. Lincoln, also, had southern relations via his wife. But Meade may have been too gentle and genteel for the kind of war the president wanted.

It was one thing to be an aggressive corps commander, but "quite a different thing," as Meade marveled earlier at Hooker's failures, to be ultimately in charge of the army. Top command weighed Meade down with the chains of fearful responsibility. "If he will," he would say of Lee, "I shall fight him at all hazards." But never after Gettysburg did he lead a major battle against him. The historian T. Harry Williams thought that his defensive victory there "ruined him as an offensive general" forever after.[46] Gettysburg was glory enough.

Close to the Potomac, Meade protested that he did not wish to "imitate" Lee's "example at Gettysburg" and demolish himself against strong defenses. Indeed some of the terrain there resembles the fields of Pickett's charge and worse.

But there the analogy stops. At Gettysburg the federals had a compact, J-shaped three-and-a-half-mile line with perhaps 90,000 men to defend it. This meant roughly 25,000 defenders per mile. The Confederate attackers had a six-mile line and 75,000 men. Then the tables turned, and mathematics gave the advantage to the would-be attackers. With reinforcements, the Union strength at Williamsport was back to 85,000 men. The defenders, however, had 43,00 effectives and a line fifteen miles long. However well engineered, that allowed fewer than 3,000 men per mile. Even though the figures are rough simplifications, the final conclusion changes not. How many Union lives was it worth to pursue fast starting on July 4, attack, albeit not necessarily frontally, and try to win the freedom of the thousands of Union prisoners Lee held as late as July 10th? How many lives to try to end the war, the possibility of which Lincoln evidently, if momentarily, managed to convince Meade of?

The perspective of the motorized warfare of another century, of course, can make the rapid pursuit of another time seem all too easy. Similarly, Civil War generals, as a rule, could not destroy an opposing army. They often seemed as crippled in victory as in defeat, not quite knowing who had won or lost. Destroying an opponent also appeared to go against the grain.

Not for a president, though, who (with his own adult son tucked safely away at college) urged the sons of others to attack. After Fredericksburg, where federal losses dwarfed those of the Confederates, the North recoiled in horror. Not Lincoln. William O. Stoddard, one of his secretaries, later recorded the commander-in-chief's startling thoughts.

> He says that if the same battle were to be fought over again, every day, through a week of days, with the same relative results, the army under Lee would be wiped out to its last man, the Army of the Potomac would still be a mighty host, the war would be over, the Confederacy gone. . . .

If this seemed terribly cold-blooded (and mistaken in the Fredericksburg context), it nonetheless pointed the way to victory and could save lives over the long run. As Mark Neely has explained, Lincoln fathomed the nature of war before the germ theory changed medicine. After all, twice as many men died from disease as did in battle. Or as Lincoln put it, "peace would be won at a smaller cost of life than it will be if the week of lost battles must be dragged out through yet another year of camps and marches, and death in hospitals. . . ." The commander-in-chief wanted a general who could face "the awful arithmetic." Meade could not. Though he would never put it this way, his son may have hit the mark best when he wrote on July 17: "Lee escaped. . . . The rebels had a very strong position, and we never could have taken it, without an immense loss of life."[47]

Lee escaped. Of course some officers party to the outcome attempted to justify it. The defense of the honor of the Army of the Potomac by its members is also understandable. Nor is it surprising to find General McClellan writing to his successor as early as July 11: "You have done all that could be done." Attacks by Meade's enemies can be discounted, too. One soldier whose opinion deserves respectful hearing is Henry J. Hunt, Meade's artillery chief, who after long reflection loyally deemed his commander *"right"* on *"all"* the controversial issues of the Gettysburg campaign but allowed for the possible exception of what took place by the banks of the Potomac.[48] Perhaps most revealing were letters home from another officer of Meade's headquarters, Major James C. Biddle.

> *July 8:* "we may be able to break up the whole of Lee's army . . . ."
>
> *July 11:* "it is impossible for me to tell whether or not a decisive battle will come off today or not. The troops are in fine condition, & I trust the rebellion is near its end. . . . Vice President Hamlin is here. . . ."
>
> *July 14* [evidently written before it grew clear that Lee's Army was gone]: "So far we have had no battle. I do not understand . . . I think . . . the rebels . . . will be compelled to attack us. . . . This is mere supposition on my part—but in no other way can I account for our not attacking. . . . the rebels are fortifying their position, and every day makes them stronger but in the same time everyone appears to be in good spirits—& Gen French said the night before last "he thought we have got them."
>
> *July 16:* "I wish we had pitched into Lee the day before he crossed the river. he had a very strong position—& five out of seven Corps Commanders opposed attacking—"[49]

By July 18 Major Biddle began to become defensive about Meade and the Army of the Potomac, yet wondered how it would be able to destroy Lee in Virginia if it had failed to do so in Pennsylvania.

In Washington Lincoln asked the same question. After the Confederates crossed the Potomac practically unmolested, his "grief and anger was something to behold,"

Noah Brooks remembered. Halleck translated that "grief and anger" into moderate diplomatic language, telling Meade of Lincoln's "great dissatisfaction." That and ten days of relentless, even insulting badgering were enough for the proud warrior. He knew in his heart that he had done his best. Washington demanded from him the "impossible," he confided to his wife, when "the proper policy" should have been being "contended" with driving Lee home to Virginia.[50] Meade tendered his resignation—as would Lee before long to his president, Jefferson Davis.

For military reasons the general of Gettysburg could not be allowed to go: no replacement seemed readily available. Lincoln could not afford to let Meade go for political reasons, either,—though this was not so much because this soldier had become such a popular hero. In fact, the unhindered Confederate withdrawal, with prisoners, booty, and all, led to much public grumbling. Even some cabinet members, in private, voiced the unfair opinion that Lee had won. But the president, who had earlier declared the bloody draw at Antietam a victory so that he could go ahead with emancipation, also badly needed Gettysburg as a great victory. Meade came with that. Building northern morale was one of Lincoln's most important tasks.

First, though, he vented his bitter frustration in private. He would never send to Meade the letter penned late on the day Lee crossed the Potomac, but it revealed clearly the president's thinking. "I am very—*very*—grateful to you for the magnificent success you gave to the cause of the country at Gettysburg," he began. Lincoln then went on to admit "such deep distress" as could not be hidden, and explained its cause. For ten days now, he saw a general who appeared bent on getting Lee back to Virginia without a battle. "You stood and let the flood run down, bridges be built, and the enemy move away at his leisure, without attacking him. . . . my dear general, I do not believe you appreciate the magnitude of the misfortune involved in Lee's escape. He was within your easy grasp, and to have closed upon him would, in connection with our other late successes, have ended the war. As it is, the war will be prolonged indefinitely." The unsent letter tells us as much about Lincoln's tears as his son's words do.[51]

Meade, too, had an unsent letter inside him: wanting to write "frankly to the President," wanting to be relieved, but he, too, stopped himself.[52] Had he not, he might have written that even in defeat Lee remained a formidable foe. The morale of Lee's men remained high, and his retreat was a model of planning and execution. The terrain was formidable along the way, and also by the Potomac, where Lee had built daunting fortifications. An assault would have taken a terrible toll, and even if successful, total destruction of a still large and experienced army seemed improbable. However much, dear Mr. President, you wish to end the war, the time has not yet come.

Nearly all of this was true, and Meade might have said more, too, and some things he did say to Halleck. The army had serious supply problems; horses, mules, and provisions were needed. Emergency troops, even experienced generals, were less than trustworthy. The command structure was broken. The leaders in Washington did not know enough about the conditions of either army. The general in the field was new to command. At Gettysburg, he won a major defensive battle, and shifting to the offensive was more than he could do. Risking defeat might risk the capital, for example, and would jeopardize the hard-won boost to Northern morale. And Meade did write by the end of July poignantly and accurately: "I . . . say to you, and through you to the President, that I have no pretensions to any superior capacity for the post he has assigned me to . . . ." And to his wife he wrote: "I can not tell you how tired and waried I am, and how I long for rest & quiet."[53]

Quickly enough, Lincoln learned more about the battle that was not fought, about some Confederate strengths and northern weaknesses. All the same, he never lost the belief that the war might have been ended in mid-July; that at Gettysburg "Gen. Meade and his noble army had expended all the skill, and toil, and blood, up to the ripe harvest, and then let the crop go to waste." The president also knew what he had to do: carry on and make the most of Gettysburg. He had known for long the power of public sentiment. Government rested on it. In the cabinet, he defended Meade and wrote to one of his generals, O. O. Howard: "A few days having passed, I am now profoundly grateful for what was done, without criticism for what was not done. Gen. Meade has my confidence as a brave and skillful officer, and a true man."[54]

Nothing testified better to Lincoln's superb ability to do what he had to do than the doggerel he scribbled in very "good humor" on July 19:

General Lee's invasion of the North written by himself:
In eighteen sixty three, with pomp,
and mighty swell,
Me and Jeff's Confederacy, went

forth to sack Phil-del,
The Yankees they got arter us, and
giv us particular hell,
And we skedaddled back again,
and didn't sack Phil-del.

No Gettysburg Address this, and so the loyal Hay, to whom Lincoln gave his scribble, managed to make it disappear for about a century. Yet later on this same doggerel day, the "tycoon" told the secretary: "Our Army held the war in the hollow of their hand & they would not close it."[55]

The president and the general never fully cleared the air. Perhaps it could not be done. As it was, what followed in the latter part of 1863 at times remarkably resembled what happened after Gettysburg. In mid-October, Lincoln once again set on edge in Washington, hoping that Meade would bring on a battle in Virginia. "How is it now?" he telegraphed one day. And the next: "What news this morning?" He and Halleck urged action, and Meade promised it, even if he would "have to attack." The opposing armies moved hither and yon but did not fight. Lincoln again tried to shame his cautious general into battle, using the same whip that he used after Gettysburg. He handwrote a letter to General Halleck, who promptly sent it on via messenger: "If Gen. Meade can now attack . . . with all the skill and courage, which he, his officers and men possess, the honor will be his if he succeeds, and the blame may be mine if he fails. Yours truly, A. Lincoln." This was old hat, and Meade was unperturbed. He quickly returned his assurances about his "intention to attack the enemy." Nothing came of it.[56]

After summoning Meade to Washington, Lincoln refused to offer detailed criticism of the Virginia operations. And Meade took no offense if (as historians Coddington and T. Harry Williams among others believe) the president recounted how the general's actions after Gettysburg reminded him of nothing so much as "an old woman trying to shoo her geese across a creek." More likely, however honest and however good a storyteller, Lincoln compared his general to an old woman in front of someone other than the butt of the anecdote. However that may be, Meade wrote to his wife that "The President was as he always is very considerate & kind." Yet Lincoln also made "very evident" his disappointment: he had wanted Lee attacked in Virginia. Meade returned to the front and would at times fancy himself as being "on the eve of a great battle." But that was all.[57]

How Lincoln really felt he already made plain to Welles a month earlier. "He said he could not learn that Meade was doing anything, or wanted to do anything. It is . . . the same old story of this Army of the Potomac. Imbecility, inefficiency—don't want to *do*—is defending the capital. . . . it is terrible, terrible, this weakness, this indifference of our Potomac generals, with such armies of good and brave men."[58]

Meade continued to maneuver in Virginia, as did his southern adversary, each hoping to get the better of the other with sureness. Such a situation never offered itself. Perhaps Meade's most interesting moment came at Mine Run on December 1, when he vacated his lines just before Lee advanced upon them. The Rebs found empty trenches, as Meade had in mid-July. Perhaps there was satisfaction in that.

But by December 1863, the general also expected to be fired for not bringing on battle. He wrote to his wife, perhaps in part because of Lincoln's irritating offers to assume blame in case of failure: "I shall write to the President . . . assuming myself all the responsibility. I feel of course greatly disappointed—a little more good fortune, and I should have met with brilliant success. As it is my conscience is clear—I did the best I could. If I had thought there was any reasonable degree of probability of success, I would have attacked. I did not think so—on the contrary believed it would result in a useless & criminal slaughter of brave men, and might result in serious disaster to the army." Meade might have written that same letter after Williamsport, too.[59]

Though he stayed in command until the spring of 1864, Meade never fought a major battle against Lee after Gettysburg. And, of course, vice versa. But all along, the Army of the Potomac substantially outnumbered the Army of Northern Virginia, and replacements were much more available to the former. In any case, the Confederacy only needed a stalemate. The United States, to save itself, and to destroy slavery, had to win. Meade seemed incapable of creating victory.

Earlier, Lincoln had burst out about his general (almost as Major Biddle had, more politely, at Meade's headquarters): "If he could not safely engage Lee at Williamsport, it seems absurd to suppose he can safely engage him now. . . ."[60] Though a specific situation induced that comment, Lincoln might have generalized the insight, were it not that he thought he had no other general to turn to. He held onto Meade until he grew convinced that Grant could be spared in the West and might make a good commander in the East, indeed a good general-in-chief. Then Lincoln made the change, letting this Illinoisian decide the fate of the Pennsylvanian. Grant kept Meade in place, though refusing him independent command or intimate confidence. Meade made a good subordinate.

When politicians and the press attacked Meade, and they did even via Congressional inquiry in the spring of 1864, Lincoln gave them little comfort. When Meade, who looked on the attacks as his "second battle of Gettysburg," demanded a court of inquiry, Lincoln refused. "The country knows that . . . you have done good service; and I believe it agrees with me that it is much better for you [to] be engaged in trying to do more, than to be diverted . . . ." The president was supportive, albeit measured, in his endorsement. Earlier, he could find no time to attend what Meade called "the grand presentation day," when the Pennsylvania Reserves gave the general a beautiful Damascus steel sword, "most chaste and artistic," but "pity . . . useless."[61] Robert Lincoln, however, did attend. Now and then the president also found occasions to make complimentary remarks about Meade. Their most frequent interactions occurred when Lincoln felt the need to review court-martial decisions in Meade's command. Their contacts were cordial. When John Wilkes Booth claimed Lincoln's life, Meade offered a eulogy. Both men died at age fifty-six, Meade in 1872, and the two shared some characteristics, among them much hard work, competence, honesty, and patriotism. But ultimately, the twain never did quite meet on common ground. Lincoln, a giant, had looked for a giant in Meade. But it was not there. Only a good soldier was.

Lincoln may never have fully appreciated the military

value of the victory at Gettysburg. He overcame his searing disappointment, however, to recognize the battle's potential primacy for building morale. This is why he went to Gettysburg on November 19, to help dedicate the soldiers' cemetery.

Edward Everett, the main speaker, understood what was at stake and could not have performed his task better as he recounted at length, and in the most glowing terms, the battle, and the pursuit of Lee that followed it. After all, the orator got a good part of his account from Meade himself. Everett tried to elevate Gettysburg to the heights of the legendary battle of Marathon. When he finished, Lincoln jumped up and "with great fervor" shook Everett's hand: "I am more than gratified, I am grateful to you."[62] He had reason to be.

Then the president's turn came. On an ever so different level than Everett, Lincoln reached for the same goal. He, too, spoke with great purpose of the larger meaning of the battle and the war. Yet some words just slipped out. "It is for us, the living," he said, "to be dedicated here to the unfinished work that they have thus far so nobly carried on." Applause interrupted his words, but he may have stumbled over the "unfinished work." At least the New York reporters wrote down the meaningless "refinished work." Others, however, heard the words right. And the nation then and since has understood these phrases, "the great task remaining before us" in their largest meaning, referring to the Civil War as a whole, surely as Lincoln wanted it.[63]

General Meade could not attend; the army required his presence. That was just as well. Had he returned to Gettysburg on November 19, in a shock of recognition he might have understood in a hurtfully personal manner the president's words about "unfinished work."

And so to a conclusion. At Gettysburg, Meade reached the most important moment of his public life. In his farewell address to the Army of the Potomac, he would mention only one battle, "the turning point of the war."[64] His biographer rightly titled his book *Meade of Gettysburg*. Lincoln, too, is inseparable from the little Pennsylvania town, though he spent but twenty-six hours there, and then spoke for little more than two minutes in an important way. And, of course, the Meade-Lincoln relationship was defined by the Gettysburg campaign.

Looking at this common moment in their lives, in the American nation's life, weighing the evidence and the arguments on all sides, I cannot reach a better conclusion than that of Sergeant Samuel Ensminger of the Army of the Potomac, a Pennsylvania draftee from Cumberland County. In private life he had been a husband, the father of a young girl and an older boy, and a prosperous distiller. At age thirty-six he could have honorably hired a substitute. Instead he answered the government's call, Lincoln's call, in the first northern draft. What education he had we do not know, but his English was peculiarly his own. Here is a sample: "Samuel Brandt is well & in joying himself well he is Surgent of thay provist gards in town it is nise growing wether here corn is 4 inches hie erley potates 6 inches hi pees nee hi flowers all in full bloom." He wrote this letter two months before Gettysburg. But if he spelled "shows" as "shoes" or "fight" as "fite," he could usually report home that "I for my part is geting along very well I would like to see you all but so it is." And he was exceedingly "Sharp," as he might have said, an acute observer of things around him.[65]

As the Army of the Potomac faced the Army of Northern Virginia before Williamsport, Ensminger wrote to his wife, Magdalena: "we are not afraid getin in a fite eney more." He also added, however, with a note of resignation: "old Lee . . . still shoes fite to our men & when thay foll on him he will slip a cross [the River] too & then thay will stand & look at him we have solgers enoughf here to ete him if thay would go rite on to him."[66]

Well, "thay" did not go "rite on to him." Sergeant Ensminger's letter was dated July 14, 1863. He did not know that what he predicted would happen, in fact, had already happened. The Confederates had already "slip across" the river. Nor could he know that his foresight had been shared by the president of the United States, though not by the commander of the Army of the Potomac. What was left then for the Sergeant Ensmingers of that Army was to "stand and look" at the other side of the wide, no longer cresting waters.

But the final word should go not to the sergeant alone, nor to Lincoln or Meade, but to the many Billy Yanks of their army, the common soldiers who sang late in 1863 their own lyrics to the sad Irish tune of "When Johnny Comes Marching Home."

> We are the boys of Potomac's ranks,
> Hurrah! Hurrah!
> We are the boys of Potomac's ranks,
> Hurrah! Hurrah!
> We are the boys of Potomac's ranks,
> We ran with McDowell, retreated with Banks,

And we'll all drink stone blind—
Johnny, fill up the bowl.

The Billy Yanks went on to sing about generals like John Pope:

He said his headquarters were in the saddle,
But Stonewall Jackson made him skedaddle.

Burnside who

. . . then he tried his luck,
But in the mud so fast got stuck.

And McClellan who

. . . was recalled, but after Antietam
Abe gave him a rest, he was too slow to beat 'em.

But the last stanza the soldiers reserved for "the old snapping turtle" himself:

Next came General Meade, a slow old plug,
Hurrah! Hurrah!
Next came General Meade, a slow old plug,
Hurrah! Hurrah!
Next came General Meade, a slow old plug,
For he let them away at Gettysburg,
And we'll all drink stone blind—
Johnny, fill up the bowl.[67]

4

Lincoln and Sherman

Michael Fellman

Close together yet far apart. *(Detail from the 1865 lithograph by Peter Kramer, "The Council of War."* Boritt Collection*)*

NOTHING COULD BE FURTHER from the truth than the common picture of the Union war effort as some sort of well-organized, well-directed juggernaut. The Union won, of course, and historians have tended to read back into the war effort some necessary evolution toward that victory. As one could infer from the argument of Richard Berringer and his co-writers of *Why the South Lost the Civil War,* perhaps the Union, which was somewhat less badly organized and motivated than the Confederacy, won by default. In any event, the mess of malice and wild disorder in Washington around the issue of civil liberties, which Mark Neely has demonstrated so vividly, ought to be extrapolated further than it has been thus far. At the center of the war effort in Washington, for lack of a bureaucratic, governmental, and martial tradition, a wide variety of officials improvised like mad and with little coordination, and managed to paste together a war effort sufficient to the day, but barely so.[1]

There was no general staff to direct overall planning for the army, and the office of the secretary of war was a jerry-built affair which frequently could not forward telegrams to the appropriate officers in the field, much less coordinate their efforts. At the supposed apex of the nation, the president, who was indeed a brilliant politician, especially in retrospect, was much unappreciated by his fellow Republicans and by his countrymen more generally during the war, prior to that martyrdom at the moment of victory which lent him his Christ-like meaning in death. As well as being disrespected by most of his nearest colleagues and by vast reaches of the public, Lincoln was, it is fair to say, no bureaucrat at all. His desk was cluttered, he had a small and inexperienced staff, his contact with the military was sporadic and scattered, he was terribly depressed and often out of touch, and he had no experience in manipulating those managerial levers which his colleagues were creating as they went along. His great strengths were his ability to articulate war goals in beautiful writing, his exquisite sense of political timing and of the politically possible, and his lack of fear of strong personalities around him. He was a man who played his hunches about other men, both cabinet officers and generals, and when they were successful, he gave them their lead, even if they despised him and worked contrary to his desires. He did bone up on military strategy and even tactics, and he certainly interfered with his losing generals, particularly in the Army of the Potomac, but this amounted more to demoralizing meddling than to coherent strategizing on Lincoln's part. In the West, where the Confederate effort had always been weaker on many grounds, Lincoln had found U. S. Grant in 1862, and soon placed him in clear command. Grant carried on with little functional reference to Washington and when he went east, he chose William Tecumseh Sherman to be his successor, which was good enough for Lincoln, who rarely intervened in Sherman's activities from then until the end of the war.

Lincoln and Sherman had almost no contact during the war. They met twice, briefly, in 1861, and did not lay eyes on one another again until March 1865. They corresponded infrequently; in their few letters, Lincoln usually would make some modest request for kindness to southern civilians, which Sherman would reject. Sherman was never called to a conference in Washington, and he rarely solicited or received military advice from that quarter, though he frequently reported his activities and requested material aid. Such decentralization typified the war in the West for both North and South. Neither the elements of the military which had been centralized in Washington, nor Lincoln exercised anything approaching on-going control.

Sherman's main contacts with Washington in 1861 were political rather than military, through his well-placed and powerful extended family. His father-in-law, Thomas Ewing, who had been an important Whig senator from Ohio and a secretary of the interior, remained an influential political figure among Republican politicians, including Lincoln, who sought out his advice and approval. Ewing's connection with Sherman was complex, because he had taken the nine-year-old "Cump" into his home following the death of the boy's father, a distinguished but impoverished Ohio judge. Cump later married Ellen Ewing, who was in effect, if not legally, his step-sister before she was his wife. This quasi-incestuous marriage also cemented a political alliance between the Ewings and the Shermans. For John Sherman, the general's younger brother, was another influential Ohio Republican, just elevated from the House to the Senate, where he took the vacancy created when Salmon T. Chase was elevated to the post of secretary of the treasury by Lincoln. With his natural pugnacity augmented by these powerful familial connections, William T. Sherman's conflicts with Lincoln would prove to be almost exclusively political in nature as well, and on those grounds he would take pubic stances at variance with the President's.

Primary among the blurred qualities of the Civil War was the way in which ostensibly military participants frequently acted as independent political operators. This was not true merely of volunteer officers, who frequently were politicians in civilian life; it was at least equally true of professional, West Point-trained officers like George McClellan and William T. Sherman. And where McClellan was a failure, Sherman was a success, and therefore gained far greater leverage in his independent political role, in which he was often antagonistic to Lincoln's leadership—theoretical subordination of the military to the commander-in-chief notwithstanding.

Notable among the many powerful Republicans in Washington who felt contempt for Lincoln until nearly the end of the war was Senator John Sherman. He wrote to his brother, the general, throughout the war, filling his ear with just how disastrous he found the president to be. In June 1862 he wrote his brother that Lincoln was "honest & patriotic, but that he so lacked "dignity, order & energy," that he "would fail in any business except pettifogging." Chase, Seward, and Stanton were John Sherman's idea of "men of ability," and the senator's contempt only hardened over the years. In May 1863 he would write his brother that he would "never cease to regret the part" he had taken in Lincoln's election, and averred that he was "willing to pay a heavy penance for this sin." He even contemplated jumping parties to ditch Lincoln. "I certainly would be glad to support a War Democrat—anybody rather than our Monkey President." By this point, John Sherman had adopted the contemptuous language of the anti-administration press, which frequently debased Lincoln as a gorilla, a black gorilla, full racism intended. In John Sherman's reading of the press for his brother later in 1863, he concluded that Lincoln would leave office in 1864 "with the reputation of an honest clown." Like many other Republicans during the long hot summer of 1864 prior to General Sherman's taking of Atlanta, John Sherman, as late as September 3, cast about for any alternative candidate, suggesting to his brother that Grant and/or Sherman would be a popular choice. The day after Atlanta fell, knowing that his brother's triumph meant the Republicans now could win with Lincoln remaining at the head of their ticket, John Sherman fell into line, writing his brother that he would support Lincoln "with all my might," despite the facts that he knew Lincoln disliked him and that he continued to feel this "singular man" lacked "dignity and energy and business capacity." McClellan was even worse, "unruly" and in the "hands of . . . traitors," but more significantly, John Sherman now sensed Lincoln would win.[2]

Far from being unique, this senator's long-term contempt and backhanded support for the president were common fare in Republican circles throughout the war. The political stakes of the conflict were enormous, and the Union effort often seemed doomed, so it should not astonish us that the embattled president was personally very unpopular, serving as a scapegoat for the frequently desperate Union condition. His office, in peacetime, had been institutionally weak as well, and although he took some harsh war measures of his own and looked the other way while others imposed draconian conditions in various places, he could hardly have been expected to have been either a widely supported or powerful figure. The lurching, rough, jerry-built Union war effort was no machine, and Lincoln was no mechanical engineer.

In the context of his brother's views and of the haphazard war effort as civil war began, it should come as no surprise that William T. Sherman was not an enthusiastic supporter of the man or of the effort, particularly at the disintegrative beginning. In his case there were additional personal as well as political reasons for distance from the oncoming conflict. During the secession crisis he had been living in Alexandria, Louisiana, where he was serving as the founding president of the military college which would evolve into Louisiana State University. He opposed disunion strongly, and he would choose the North if it came to that, but he was extremely conservative on all other social and political issues, favoring the retention of slavery, opposing the Republican party, despising American democracy, even fantasizing about a military dictatorship of the sort that Louis Napoleon had imposed on France.[3]

Following his resignation in Louisiana on January 18, 1861, shortly before that state seceded from the Union, Sherman steamed up the Mississippi River toward Ohio, and began looking for a job. Soon after he arrived back in Lancaster, Sherman received an offer from his old banking partner, Henry S. Turner, of the presidency of the St. Louis Railroad Company, a rather inflated title for the Fifth Street horse-car line. This respectable, if modestly paying ($50 per week) job would be, he hoped, a good entrée into the business community of booming St. Louis. However, the same mail that brought him Turner's job offer brought him a letter from his brother, urging him to come to Washington immediately to discuss a major military role directly with the new president.

There is no trustworthy account of this brief meeting of Lincoln and the Sherman brothers, but even Sherman's self-interested recollection of it, written fifteen years later in his *Memoirs,* indicates that it was a disaster. John introduced his brother as fresh from Louisiana. "Ah, how are they doing down there?" Sherman recalls Lincoln as having asked, probably in that bantering manner in which Lincoln handled most stressful situations, much to the discomfort and even anger of his less than ironic visitors. Sherman told Lincoln that the southerners were, "getting along swimmingly—they are preparing for war," to which Lincoln, who thought that the southerners were bluffing and doubted that war was imminent, replied, "Oh well, I guess we'll manage to keep house." Sherman later recalled that he had been stunned by this jocular dismissal of what he was absolutely certain was a national military emergency. "I was silenced, said no more, and we soon left." Walking out of the White House, Sherman exploded to John, "damning politicians . . . [saying] you have got things in a hell of a fix and you may get them out as best you can." The political volcano was soon to erupt, but he was off to St. Louis to take care of his family and would have, "no more to do" with blundering politicians and their squalid messes. John begged him to be patient, but he said he would not be so and would return to the Midwest immediately.[4]

Sherman had felt personally insulted and rejected by the leader of that stupid political caste which had caused what he was certain was an incipient war. Lincoln in fact may have attempted to placate Sherman later in their interview, but Sherman had been so infuriated that not only had he remained silent while with the president; he also could not recall in later years anything else which Lincoln may have said. And then he decided, as was his wont, impetuously and impulsively, to sit out the war as a businessman. Off to St. Louis he went to be the streetcar-line president, firing workers to improve the profitability of his company, while Mrs. Ellen Sherman also settled in by buying expensive furniture and a first-class Brussels carpet to furnish their rented house on Locust Street. During these same springtime weeks, the opening gambits of the war increased in pace, moves which would lead to the explosion at Fort Sumter on April 12. While these momentous events were unfolding, John worked hard to procure the chief clerkship of the War Department for his brother. Immediately after Fort Sumter, the energetic John conveyed a War Department offer to Sherman that he become the major-general in charge of all Ohio volunteers, and the much distrusted Frank Blair offered Sherman the parallel post for Missouri. He turned down both these chances, out of a deep funk, at once emotional and ideological.

Sherman believed that his having left Louisiana was in itself a unique declaration of loyalty by a professional army man, during a period of trial when so many others had deserted to the South. "I am the only northern man who has declared fidelity to the Union in opposition to the modern anarchist doctrine . . . of secession," he wrote John, more in heat than in accuracy. Republican politicians, including the president, "in shameful neglect & pusillanimity," had only wanted Republican hacks, not West Pointers, to defend the Union. While the South "try to attract" the best military men (soldiers just like William T. Sherman), "the north don't care a damn." Lincoln was incompetent. "I have no confidence in the head and advisors." Lincoln had rejected him to his face, rejected him right out of hand. "Had Lincoln intimated to me any word of encouragement, I would have waited" in Washington for a command. He would not be drawn into the "muddle" of the first call for three-month volunteers—"I like not the class from which they are exclusively drawn [i.e. Republicans]—a class which inevitably, "will be defeated and dropped by Lincoln like a hot potato." Nor would he return to Ohio to head the volunteer effort in that state, for "no man is a prophet in his own land." The current war effort was unworthy of him, as were the American people, and he would bide his time, "until the [political] leaders will be cast aside," and a new set of men will arise to create a real military government—the fantasy of imminent dictatorship he had been polishing for quite some time.[5]

Sherman climaxed this emotionally charged package of

resentments on April 8, when he wrote to Montgomery Blair, the postmaster general, rejecting the War Department chief clerkship, and concluding, "I thank you for the compliment . . . and assure you that I wish the Administration all success in its almost impossible task of governing this distracted and anarchical people." John let his brother know that this letter had given great offense when it was passed around in Washington, and that several of the cabinet had concluded that Sherman would turn traitor, as had so many other Union officers, northerners included. As it happened, at about this time, Sherman was approached by at least one Louisiana Confederate as a potential military recruit, to whom he apparently did not reply.[6]

After several weeks of rapidly escalating public events, and considerable pressure from his family, the shame and isolation of failing to participate in the war effort became insupportable, and so, on May 9, Sherman volunteered his services to the secretary of war. Fortuitously, at this moment, the government created several new regular army regiments, and Sherman was offered the colonelcy of the 13th Infantry, which he accepted. Where he would turn down a major-generalship of a corps of volunteers, tainted as such units were for him by the low, democratic politics which characterized their recruitment and choice of officers, this West Point elitist would accept the lesser position of colonel in charge of a regiment of professional troops, which were to him above politics.

The overwhelmingly volunteer and democratic army he was joining in Washington that summer was a pell-mell ingathering. Americans had been willing to maintain only a tiny army in peacetime; and so they had to improvise, equip, train, and send into war an immensely complex social organization starting from scratch. For several weeks in Washington, Sherman was deputized the job of inspecting and organizing raw troops from all over the North. He did not join his brigade of one Wisconsin and three New York City regiments—an untrained "rabble," he frequently called them—until July 5, and he had the opportunity to drill them only a handful of times before they joined the march on July 16, on to Richmond.

Bull Run, the battle which ensued, was a Union disaster. After advancing in relatively good order for most of the day, in mid-afternoon, the union army broke and fled back to Washington. No worse nor better than other Union troops, Sherman's brigade exemplified this initial attack and later rout. Back in the fortifications outside Washington, Sherman wrote to the army authorities that his brigade, "in common with our whole army [had] sustained a terrible defeat, and has degenerated into an armed mob." As for himself, Sherman wrote to his wife, "I am absolutely disgraced now," so much so that he wished to "sneak into some quiet corner," following this "mortification of retreat, rout, confusion, and now abandonment by whole regiments."[7]

The military leadership did not agree with Sherman's private self-devaluation, and gave him a de facto promotion to brigadier-general, adding four more regiments to the four already under his command. To his way of thinking, this affirmation just added to his problems, as he had not received real troops, but "volunteers called by courtesy soldiers," as he wrote to Ellen. To John he added that three of his eight regiments were in a state of mutiny, half of them, "clamorous for discharge on the most frivolous pretexts," over one hundred of them prisoners in the hold of a nearby man of war, and the one company of regulars among this rabble, "all ready with shotted guns to fire on our troops."[8]

In his 1875 *Memoirs,* Sherman wrote that at just this juncture a captain from one of his New York regiments, an Irish lawyer, announced to Sherman, while standing among a crowd of his fellow Irishmen, that his three-month term was up and he was going home, to which Sherman recalled as having barked, "If you attempt to leave without orders, it will be mutiny and I will shoot you like a dog," reaching his hand into his overcoat as if to draw a pistol. The Captain looked at Sherman hard, and backed down. Later that day, President Lincoln and Secretary of State Seward came riding by in a carriage and invited Sherman to join them while they rallied the troops. Inside the encampment, the carriage stopped and the mutinous captain approached it, announcing, "Mr. President, I have a cause of grievance. This morning I went to Colonel Sherman, and he threatened to shoot me." After a moment's hesitation, Lincoln leaned down to the captain and said in a stage-whisper, "Well, if I were you, and he threatened to shoot, I would not trust him, for I believe he would do it." The men all roared, laughing at the captain, and the carriage drove on, Sherman quickly explaining himself. Lincoln reassured him, "Of course I didn't know anything about it, but I thought you knew your own business best."[9]

Lincoln gave every appearance of approving of him, but following the rout at Bull Run, such scenes as that created in front of his men by the Irish lawyer hit Sherman hard and made him feel isolated and shaky. While in this condition, in mid-August, Sherman was asked by Robert Anderson, the ailing and aging hero of Fort Sumter, to join him as his second in command in Kentucky, where Anderson had been named head of the newly created Army of the Cumberland. Sherman agreed to his new appointment, but only reluctantly, and only after explaining to the president his "extreme desire to serve in a subordinate capacity, and in no event to be left in a superior command." Lincoln readily agreed to this odd request, joking that his usual problem was in finding places for generals who demanded to be at the heads of armies. To Ellen, Sherman wrote about his new position that, "not till I see daylight ahead do I want to lead," thereby expressing his lack of confidence not only in the Union war effort and in Lincoln, but also in himself.[10]

Despite his wishes, Sherman soon followed the sickly Anderson into command in Kentucky, a role he was mentally incapable of fulfilling. Sherman saw hidden enemies everywhere in Kentucky. He became convinced that such immense numbers of enemy troops were gathering in secret that he would need an army of 200,000 even to hold the state. This would have been twice the size of the Army of the Potomac. Sherman may even have feared that he would have to abandon Louisville and retreat across the Ohio River. Rather than keeping such delusionary fears to himself, Sherman shared them both with newspapermen and with Secretary of War Simon Cameron, when Cameron visited him in Kentucky. Within a month of assuming command, Sherman had fallen into a deep, clinical depression, sleeping and eating little, smoking cigars, talking and probably drinking obsessively, pacing the corridors of his hotel at night, alarming his staff with his loss of human contact and his increasingly compulsive habits. On November 9, 1861, Sherman was relieved of his command in disgrace, a fall from power soon broadcast by the newspapers across the nation, which all shouted on December 11 that Sherman had gone insane.

Having been wired by his staff to come fetch him in Louisville, Ellen Ewing Sherman took a hard look at her husband's condition and brought him back to Lancaster for rest and reassurance. With the cooperation of the tactful Henry W. Hallek, Sherman's superior, Ellen went on the attack to rescue her husband's career and his reputation, organizing and directing the powerful Ewing/Sherman forces in Washington, and personally taking on the job of getting right to the top—to Abraham Lincoln. In her opening salvo, Ellen Ewing Sherman wrote a powerful letter to the president on her husband's part. "Being of a nervous temperament," she wrote Lincoln, Sherman had shrunk from "the responsibility of his position [in Kentucky], which supplied no adequate means of defense." His perfectly reasonable requests for men and material had been ignored in Washington; only his request to be relieved was complied with, and that all too "readily," if "cooly." A week after he had left St. Louis for his furlough in Lancaster, "conspirators" in the military and the press, of whom she named Adjutant-General Lorenzo Thomas, had "had time to arrange their plans," for the simultaneous publication across the Union that Sherman was insane. "No official contradiction has yet appeared and no official act has reinstated him," she wrote Lincoln. "Will you not defend him from the enemies who have combined against him?" she implored Lincoln. She asked him to send for the general, or make "some mark of confidence," to relieve Sherman from "the suspicions now resting on him."[11]

In Washington, Ellen Sherman's letter met with a sympathetic and almost instantaneous response from Lincoln, news of which Ellen eagerly passed along to her husband. General Hugh Boyle Ewing, Ellen's eldest brother, also wrote Sherman that he had been making the rounds of the army brass in Washington, who had assured him of Sherman's high standing, adding that a day or two before, Lincoln had praised Sherman's "talent & conduct," to a large group of officers. Lincoln's response demonstrated that Ellen's letter had its intended effect and in just the way she had hoped. She understood in her bones the essentially political nature of the Union war effort, and knew how to activate the most powerful levers. Lincoln's word would carry enormous weight with the toadies on the army staff in Washington, and their subsequent opinions would radiate out into the field. Her family gave her great political influence, of which her husband would be the beneficiary, receiving the benefit of the doubt and the second chance not given to politically less well-connected officers. Ellen Ewing Sherman assured her husband that subsequent to her correspondence with Lincoln, "John Sherman will attend to your interests in Washington," as would Thomas Ewing, Jr., a powerhouse in Kansas Republican circles "who has got great influence," and Thomas Ewing, Sr., the Old Whig wheelhorse who was "great friends just now," with the president, who like Ewing had emigrated into the Republican party from a Whig background.[12]

Ellen Sherman capped off her husband's rehabilitation with a personal call on the president. By the time of this lengthy interview on January 29, 1862, Lincoln was well prepared. He had every practical reason to assuage this powerful Ohio (and Kansas) Republican clan, and he knew just how to talk to the aggrieved Ellen Ewing Sherman. As she wrote Sherman the evening after the meeting, Lincoln had greeted her with the "highest praise" for General Sherman. Lincoln said he and Seward had been "strongly impressed" with Sherman when they had shared that carriage in Washington right after Bull Run, that he personally had promoted Sherman to brigadier-general because of this positive impression, even before the Ohio congressional delegation had forwarded Sherman's name for promotion, and that he had felt *"safer"* after Sherman had taken command in Kentucky "than *before*." "He said he wanted *you* to know," Ellen wrote, that he had the "highest & most generous feelings towards you," and that "recent reports,"—a delicate phrase for Lincoln to have used about the newspaper damnations—"were unfounded [and] that your abilities would soon merit promotion." Ellen Sherman and her father were delighted with this "most satisfactory interview. . . . A little time will wear away this slander and then you shall stand higher than ever," she closed her report to her husband of her assault on Lincoln.[13]

Lincoln had stuck to William T. Sherman for political rather than military reasons, in large part because this general had had a powerful apparatus in place to defend that flank for him. After all, to this point, Sherman had been another in a string of military failures, and Lincoln had received no evidence that Sherman was other than highly erratic in behavior and temperamentally very strange. However, in one of those extraordinary reversals so common in war, when he was back in the saddle in the West, Sherman almost immediately underwent a triumphant military resurrection at the battle of Shiloh, in April 1862. Sherman was in the thick of fighting; he was wounded in the hand, and had at least two horses shot from under him; he was cool and forceful, receiving special mention from Grant, and cheers from his men. This triumph, which changed Sherman's mood almost over night from bitter self-loathing to triumphal self-confidence, delighted Lincoln, and also tightened Sherman's ties with Grant. His star would continue to rise as the lesser binary twin to Grant's. However, if he ever felt gratitude to Lincoln for his second chance, he never wrote about it, and he was a man who always wrote about everything that crossed his mind. Indeed he would repay the president with maintenance of the great distance between them, and, on the one notable issue over which they had their most intense interchanges, the employment of black regiments, Sherman would rebuff Lincoln with rudeness and active insubordination.

In the racial debate as of 1863, Lincoln was toward the liberal end of the spectrum. He had always abhorred slavery on moral grounds, he was quite open to the Radical wing of his party, many of whom were prominent both in the Congress and in his cabinet, for the Radicals had the most energy and the clearest ideology and program in the nation. His greatest arguments concerning emancipation and the use of black troops were not with the Radicals but with the more conservative members of his party and of the northern electorate more generally. When he tried to persuade these conservatives of the need for a new deal on race, some of the arguments he employed were practical in nature, as he thought these might be more effective than more ideologically and morally based modes of persuasion. As he wrote to one conservative northern politician on August 26, 1863, "I thought that whatever negroes can be got to do as soldiers, leaves just so much less for white soldiers to do, in saving the Union," an opinion he believed most of his commanders shared, even conservative ones. But Lincoln went on in this letter to link the practical with the ideal, for he argued, "negroes, like other people act upon motives . . . If they stake their lives for us, they must be prompted by the strongest motive—even the promise of freedom. And the promise being made, must be kept." Going well beyond merely practical arguments, Lincoln thus linked *white* honor both to emancipation and to the concomitant use of black troops. He also comprehended *black* honor, for he understood the liberation of the spirit which military participation would bring to the freedmen. "There will be some black men who can remember that with silent tongue, and clenched teeth, and steady eye, and well-poised bayonet, they have helped mankind on to this great consummation" of freedom. Lincoln also insisted that white men who opposed the use of black troops, by striving to hinder this consummation of freedom, "with malignant heart and deceitful speech," would be acting out of a spiritual dishonor of which they would be ashamed in future years.[14]

Perhaps most, but certainly not all Union commanders agreed with Lincoln, either on practical or on idealistic grounds. Of all the leading Union generals, Sherman was by far the most outspoken in his resistance to this revolution in racial policy, the most overtly racist in his opposition, and the most openly insubordinate to civilian dictates, from those issued by the president on down. When Lorenzo Thomas began the active recruitment of black troops in Mississippi in April 1863, during the campaign against Vicksburg, Sherman made his feelings perfectly clear. In a letter to Ellen he wrote, "I would prefer to have this a white man's war and provide for the negroes after the time has passed. . . . With my opinion of negroes and my experience, yea prejudice, I cannot trust them yet. Time may change this but I cannot bring myself to trust negroes with arms in positions of danger and trust." Time and the spread of Union recruitment among blacks only served to deepen Sherman's prejudices. As he wrote to Henry Halleck late in 1863, Sherman believed as an abstract principle that the southerners would only "finally submit" to fellow white men who displayed courage and skill equal to theirs, superior resources and "tenacity of purpose," and a clear and single purpose—"to sustain a government capable of vindicating its just and rightful authority, independent of niggers, cotton, money, or any earthly interest."[15]

In tandem with the prejudices of most of his soldiers, Sherman wanted to keep his troops free from the contamination they believed Negroes would bring. When Lorenzo Thomas came west and addressed his men, informing them they would have to adjust to the presence of black troops, Sherman followed, telling his men—"who look to me more than anybody on earth"—that he hoped that if the government did make use of black troops, "they should be used for some side purpose & not be brigaded with white men." "I won't trust niggers to fight yet," he wrote John Sherman, adding that he did not oppose taking them from the enemy, and finding other uses than combat for them.[16] He was perfectly willing to use blacks as laborers and in "pioneer brigades," to dig the trenches, build the forts, chop the wood and haul the water, all in aid to the real, white soldiers.

The more actively the government pursued its recruitment of black troops, the more urgently and angrily Sherman resisted the policy. In the spring of 1864, as Sherman was organizing his campaign against Atlanta, Lorenzo Thomas came west again to coordinate the use of agents from northern states come to recruit southerners, primarily blacks, into the army as one means of filling up draft calls made on those northern states. "My special duties here are to organize colored troops, and I expect full cooperation on the part of all military commanders to enable me to execute these special orders of the Secretary of War [in which] the President has taken an interest," Thomas wrote to Sherman very pointedly from Natchez. Instead of cooperating, Sherman issued special orders that recruiting officers would not be allowed "to enlist as soldiers any negroes who are profitably employed" by the army, that his staff officers "will refuse to release him from his employment by virtue of a supposed enlistment as a soldier, and that any recruitment officer . . . who interferes with the necessary gangs of hired negroes," would be arrested and if necessary imprisoned.[17]

This was insubordination. On the other hand, Sherman was coordinating a most promising campaign against Joseph Johnston's army on the way to Atlanta, and so rather than punish him, the authorities in Washington, without quite conceding that he had the authorization to override their clear orders, tried to sweet-talk Sherman. Lorenzo Thomas wrote Sherman that the secretary of war wished, "to express his strong desire" (rather than ordering him), that Sherman cooperate in the recruitment of black troops at least by sending them on to Thomas, now in Nashville. "I have seen your recent order respecting the enlistment of negroes," Thomas continued, "the practical working of which, it seems to me, will stop almost altogether recruiting in your army. I know not under what circumstances it was issued, but the imprisonment of officers for disobedience seems to me a harsh measure." Reading this rather plaintive letter, where he might have anticipated an angry rebuke accompanyed by a renewed order, Sherman knew he had the upper hand. "I must have labor and a large quantity of it," he insisted. "I confess I would prefer 300 negroes armed with spades and axes than 1000 as soldiers. Still I have no objection to the enlistment of negroes if my working parties are not interfered with." And then, defiantly reasserting this authority over Union recruiting agents on his turf by reaffirming rather than rescinding his offensive order, Sherman concluded that if his Negro working parties "are interfered with," by recruiting officers, "I must put a summary stop to it."

Sherman was not retreating but stonewalling. Indeed, in the guise of cooperation he soon wrote again to Thomas proposing the regularization of pioneer brigades as an alternative to the Union policy of black troops. After all, he concluded, "the great mass of our soldiers must be of the white race, and the black troops should be for some years be used with caution and with due regard to the prejudice of the races." Pressing forward to Atlanta, Sherman continued to protest black recruitment with increasing anger. "I must express my opinion that [such recruitment] is the height of folly," he wrote to Henry Halleck on July 14, in a letter he must have intended to reach Stanton and Lincoln. As for the recruiting officers, "I will not have a set of fellows here hanging about on any such pretenses." Sherman was having his way.[18]

Four days later, the president himself replied to Sherman's string of insubordinate dispatches. As was his wont, Lincoln wrapped firmness in the cloth of tact. He was unwilling "to restrain or modify the law" on black recruitment, "further than actual necessity may require," he wrote Sherman, adding that, "to be candid, I was for the passage of the law myself." He had not apprehended that the law would prove so inconvenient to armies in the field, "as you now cause me to fear." Despite Lincoln's acknowledgment of Sherman's sensibilities, Lincoln got to the point quite clearly, if with civility, reminding Sherman of the constitutional subordination of the military to the president as commander-in-chief. "I still hope for advantage from the law; and being a law, it must be treated as such by all of us. We here will do what we consistently can to save you from difficulties arising out of it. May I ask, therefore, that you will give your hearty cooperation?"[19]

Writing to others who opposed his policy at about this time, Lincoln was far more direct and forceful about the utility of black troops. To Isaac M. Schermerhorn, a Buffalo New York, conservative Unionist, Lincoln wrote on September 12, 1864, "Any different policy in regard to the colored man [than black recruitment] deprives us of his help, and this is more than we can bear . . . . This is not a question of sentiment or taste, but one of physical force which can be measured and estimated as [can] horse-power and

Steampower . . . . Keep it and you can save the Union. Throw it away, and the Union goes with it."[20] Lincoln personally believed in the moral power of black troops as well, but he was willing to argue on the less elevated grounds of practical materialism if that would help convert racist Unionists like Schermerhorn. Lincoln was even more muted ideologically with Sherman, addressing his most recalcitrant general not on the moral or even the practical merits of the issue, but by appealing to Sherman's most abstract and general belief in law and constitutionalism. He asked not for conversion of belief but for mere acquiescence in a policy which Lincoln understood Sherman could not accept in his heart.

In reply to Lincoln, Sherman telegraphed that of course he was a great believer in the due subordination of the military to civilian control, in principle: "I have the highest veneration for the law," he wrote Lincoln, "and will respect it always, however it conflicts with my opinion of its propriety." Conceding the abstract principle, Sherman did not even address enforcement of the law at hand, instead reminding Lincoln, under a rather thin veneer of politeness, exactly who was going to win the battle which would save whose political skin. "When I have taken Atlanta and can sit down in some peace I will convey by letter a fuller expression of my views." Grant was bogged down and suffering frightful losses in Virginia; Lincoln's political stock was never lower; only Sherman was advancing in a way which could reverse Union fortunes in general and Lincoln's in particular. Lincoln knew all that and Sherman knew it too. At this level, war was politics by other means, and Sherman fully understood that his hand was stronger than Lincoln's, at least for the foreseeable future. He had thrown Lincoln's request for cooperation right back in his face.[21]

Although he claimed he lacked the time and peace to respond fully to Lincoln's order phrased as a request, Sherman did have time the following week to write a full reaffirmation of his independent political policy to John A. Spooner, the recruitment agent for Massachusetts, even while offering Spooner a pass within his lines to allow his pursuit of black recruits in the South, an apparent concession to Lincoln's request. Among the reasons he opposed Spooner's project, he wrote him quite candidly, was that, "the negro is in a transition state, and not equal of the white man." While claiming that he had conducted far more Negroes to safety behind Union lines than had any other Union general, he asserted that, "I prefer some negroes as pioneers, teamsters, cooks and servants; others gradually to experiment in the art of the soldier, beginning with the duties of local garrison," and inferentially, none at all serving with his fighting forces.[22]

When it leaked into the press a few weeks later, as Sherman surely anticipated it would when he circulated it to many of his officers as well as to Spooner, this defiant letter made a sensation all over the North. "I never thought my nigger letter would get into the press," Sherman pretended to a St. Louis friend, but since it had made a splash, Sherman continued, "I lay low. I like niggers well enough as niggers, but when fools and idiots"—those professing "nigger sympathy," as he put it in another letter—"try and make niggers better than ourselves, I have an opinion."[23]

Sherman used only a slightly more refined tone when, after the fall of Atlanta, as he said he would, he wrote

Henry Halleck replying in full to Lincoln's abjuration to change on the issue of black recruitment. "I hope anything I have said or done will not be construed unfriendly to Mr. Lincoln or Stanton," Sherman wrote, perhaps with irony intended, or perhaps without. "That negro letter of mine I never designed for publication," he now wrote, without disowning one whit of its contents. "I am honest in my belief that it is not fair to our men to count negroes as equals. Cannot we at this day drop theories and be reasonable men?" Sherman asked, meaning that innovative assertions of black equality be dropped for the customs of racism. "Is not a negro as good as a white man to stop a bullet?" Sherman wrote he often had been asked. "Yes," he now answered, "and a sand bag is better; but can a negro do our skirmishing and picket duty? Can they improvise roads, bridges, sorties, flank movements, etc., like a white man? I say no." Blacks were simply inferior as a race. "Soldiers must do many things without orders from their own sense . . . . Negroes are not equal to this." Sherman wrote Halleck that he intended to march in the future as in the past at the head of a lily white army. "I have gone steadily, firmly, and confidentially along, and I could not have done it with black troops, but with my old troops I have never a waver of doubt, and that very confidence begets success."[24]

Despite his continued insubordination on the use of black troops, it is an ironic truth that Sherman's capture of Atlanta in September 1864 did indeed have the political impact Lincoln had hoped it would, for that victory immediately reversed public opinion on the progress of the war and led directly to Lincoln's substantial electoral triumph two months later. However far out of control he was in politics and in the military policies which affected his army, at this critical juncture, Sherman was indeed, by his successful generalship, the chief military figure responsible for the Republican continuation until victory of the war against slavery. Both Lincoln and Sherman had understood the bigger political and moral picture, if from radically different perspectives, and both men wanted to defeat the Confederacy. Sherman had served, in absolutely indispensable ways, both Lincoln's immediate political needs and his larger war aims, even though he disagreed with those socially revolutionary goals. For his part, Lincoln showed great coolness in his ability to separate the overall political impact Sherman's military advance was making from the nasty political agenda Sherman was pursuing on very important if not, finally, strategically crucial racial issues.

After his re-election, Lincoln stubbornly returned to his conflict over army racial policy with Sherman, demonstrating anew just how strongly he felt about the use of black troops. But only after Sherman had captured Savannah did the Lincoln government succeed by fiat in attaching a black infantry regiment to Sherman's command, by shipping one in by sea from the East. What he was forced to accept, the equally stubborn Sherman subverted. Simply refusing to integrate these trained troops into one of his brigades of regular infantry, he stripped them of their arms and converted them into laborers, teamsters, and servants. Picking up on the spirit of racial contempt which Sherman expressed in these actions and in his racial policy pronouncements, Sherman's soldiers brutally harassed these black recruits, rioting against them, killing two or three and wounding many more. As one Ohio soldier wrote home, these few hundred black soldiers, "were taught to know their place & behave civilly." Openly encouraged by their commander to maintain their prejudices rather than to reform them, the overwhelming tenor of this army was to reconfirm their deep belief that they were making a white man's war.[25]

The administration effort to bring around the insubordinate hero of Atlanta and the march to the sea reached its unexpected denouement on the morning of January 11, 1865, when no less than the Secretary of War disembarked from a federal revenue cutter at the port of Savannah. Stanton urged Sherman to come up with some plan that would "meet with the pressing necessities of the case," of the existence of vast numbers of freed slaves in the South. At least that is the way Sherman reconstructed in his *Memoirs* what had followed this meeting. Sherman recalled that he had then sat down and drafted his Special Field Orders #15, which he issued after Stanton had edited them carefully. Other historians have stressed Stanton's role in the authorship and that of congressional Radicals. Whatever their exact genesis, these orders were an extremely radical proposal for redistribution of land confiscated from slaveholders to the newly freed slaves. In the sea islands of South Carolina, and for thirty miles up the rivers from the sea in South Carolina and parts of Florida and Georgia, "abandoned" plantations (from which the owners had fled on the approach of Union troops) were to be distributed in plots of "not more than forty acres of tillable ground," to black heads of families, each of whom was to be given "a possessory title in writing."[26] Beyond emancipation itself, this was the single most revolutionary act in race relations during the Civil War. How extraordinarily ironic it appears that this reactionary racist would father this extreme measure. Many historians have even insisted in disbelief that Stanton was the real author of this plan, and that Sherman went along with him, not fully realizing where he was being led.

If Sherman's Special Field Orders #15 were pleasing to blacks and Radical Republicans, that came as a bonus which made the orders even more useful to him than they otherwise would have been, as they would get both these sets of antagonists off his back. He would be rid of the blacks who followed him in this train, divesting himself of them as he would his excess mules and horses. At the same time, and of primary importance to him, he would, through this measure, keep young black males out of his army. If others took that as beneficence, so much the better, as he could then lay claim to the title of friend of the freedmen. And in addition, land confiscation would be a major blow to the planter aristocracy who were running the war for the enemy. He wanted to smite his enemies in every possible way, and land redistribution would grievously injure the morale and material fortunes of his enemies, demonstrating their powerlessness before their enemy, and humiliating them publicly with the insulting image of social pollution contained in the very thought that their ex-slaves would own their land. They would further be reduced toward nullities, having sacrificed their rights to their own property. In absentia, the southern gentry would be able only to shriek in fury.

With this highly ironic act of legerdemain, Sherman concluded his history of reactionary insubordination to Lincoln's policy on black troops by reaching to Lincoln's left to form an improbable political alliance on this one issue with Radical Republicans. He had maintained an independent racial policy throughout his period of major command, acting opposite to Lincoln's wishes. As long as he was winning battles, Lincoln neither punished nor curtailed Sherman, but theirs was a strained alliance of political/military associates, rather than a clear war effort coordinated and directed from the top down.

In general, Sherman and Lincoln shared only one goal, the defeat of the South and the preservation of the Union. In the articulation of even that great goal, however, the two men walked on opposite sides of the road. Lincoln, though intent on southern capitulation, maintained a tone of fellow feeling and reconciliation throughout the war, while Sherman became the most articulate Union propagandist of destruction and punishment for the Confederacy. Nowhere can one discern this difference more readily than in a comparison of Lincoln's biblical, anti-slavery second inaugural address of March 4, 1865, with Sherman's equally powerful jeremiad against the southerners delivered seven months earlier to the citizens of the newly fallen Atlanta.

Lincoln chose as his text the words of St. Matthew, "Woe unto the world because of offences! for it must needs be that offences come; but woe to that man by whom the offence cometh!" The offence was the sin of slavery, so terrible a sin that it could only be ended by war. Yet Lincoln did not blame the slaveholders alone for the national sin, but included all Americans as those who justly were paying the terrible price for having permited the prolonged existence of this sin. "If God wills that [war] continue, until all the wealth piled by the bond-man's two hundred and fifty years of unrequited toil shall be sunk, and until every drop of blood drawn with the lash, shall be paid by another drawn with the sword, as was said three thousand years ago, so still it must be said, [as it was said in Psalms] 'the judgments of the Lord are true and righteous altogether.' " Counter-balanced with this promise to lead the nation through war for as long as it might take, and as deep as it might go, Lincoln prayed that "this mighty scourge of war may speedily pass away." He then held out the olive branch, from a fellow sufferer of the war to his suffering southern brethren, promising "With malice toward none; with charity toward all . . . to do all which may achieve and cherish a just and lasting peace." As here, stern determination in the war and forgiveness as the final lesson of it had characterized Lincoln's sense of war purpose from early in the conflict. His hand held the olive branch in final benediction to the fallen of both sections, all of whom had fought serving the same God.[27]

In contrast to Lincoln's guilty as well as angry statement of war goals as enunciated in his last great address, in contrast to Lincoln's spirit of forgiveness wrapped inside the iron words of St. Matthew, Sherman articulated his war goals in language similar to Lincoln's, though with a different and far more destructive message. He had intended to induce terror in order to produce a defeat which would be spiritual as well as military. As he wrote ten years after the war, "My aim then was, to whip the rebels, to humble their pride, to follow them to their inmost recesses, and make them fear and dread us."[28]

Sherman's fullest enunciation of his aims as a warrior of terror came immediately after Atlanta fell to his army, when he initiated a plan to expel all civilians from the city, something he had done on a much smaller scale before, but which at this level amounted to one of the most extreme actions yet taken by any general in the war. Though he had a military argument for getting civilians out of the way of his army—in order to use Atlanta as one big military rail depot—his inner purpose was to strike terror into southern hearts. He would be the willing swordsman of barbarity and cruelty, taking on to himself a role neither Lincoln nor any other Union general could bear.

In the grip of this warrior purpose, Sherman wrote John Hood, the Confederate commander to his south, offering a ten-day truce, during which he would ship every Atlanta civilian through Union lines to the Confederacy. Hood agreed to the truce, realizing that Sherman really had not given him any alternative. But Hood added, in a letter he wrote to Sherman but which he intended primarily for the southern press, "The unprecedented measure you propose transcends, in studied and ingenious cruelty, all acts . . . in the dark history of war. In the name of God and humanity I protest." Sherman replied in a letter he intended for the northern press and not just for Hood, "If we must be enemies, let us be men and fight it out . . . and not deal in such hypocritical appeals to God and Humanity. God will judge in due time," but military men, real men, will act out of pure force rather than hiding in effeminate appeals to higher moral power. In any event, Hood's appeal to a "just God" was "sacrilegious," Sherman insisted, for it was the South which had "plunged a nation into war, dark and cruel war, who dared and badgered us to battle." Having created the war, the South would now experience it.

When the mayor and two city councilmen of Atlanta wrote Sherman, also protesting the expulsion order, which, they were certain, would lead to unimaginable "woe, horrors and suffering" on the part of the expelled civilians, Sherman replied in grim glee, "You might as well appeal against the thunder-storm as against these terrible hardships. They are inevitable, and the only way the people of Atlanta can hope once more to live in peace and quiet at home is to stop the war, which can alone be done by admitting that it began in error and is perpetuated in pride." War was a natural catastrophe which the stiff-necked secessionists had brought down on themselves. War had now visited them—as by extension it would visit all southerners. They had to cry out in defeat before the war could cease. "You cannot qualify war in harsher terms than I will. War is cruelty and you cannot refine it," thundered Sherman the Jeremiah, in King James cadences every bit as powerful as Lincoln's, "and those who brought war into our country deserve all the curses and maledictions a people can pour out."

Following up on this expulsion of the citizens of Atlanta, Sherman deepened his terrorizing strategy against the people of the South by planning his expedition through the Georgia heartland to Savannah and the sea. While making his plans, Sherman often repeated and sharpened his sense of harsh and scourging purpose. To General George Thomas he wrote on October 29, "I propose to demonstrate the vulnerability of the South, and make its inhabitants feel that war and individual ruin are synonymous terms." He was going to visit physical destruction and mental humiliation on each southerner his army met, one by one. Pulverizing each southern atom he confronted would, taken inductively and collectively, demonstrate the meaning of southern defeat to southerners. On November 8 in a letter to Grant which is now of legendary standing, Sherman wrote, "If we can march a well-appointed army right through [Jefferson Davis's] territory, it is a demonstration to the world, foreign and domestic, that we have a power which Davis cannot resist. This may not be war, but rather statesmanship." It would be the most explicit public demonstration of the most elemental political logic. "If the North can march an army right through the South it is proof positive that the North can prevail in this contest, leaving only open the question of its willingness to use that power." Only in that last phrase was Sherman disingenuous or modest or misguided: without any doubt he intended to demonstrate quite clearly that he was not merely willing but eager to use that power, for just as long as it took.[29]

This was war as politics stripped bare. Sherman was now in a position to do as he willed, which did not mean that he would do everything in practice. Indeed, even in his rage, he calculated and recalculated the effects of his means during his *March to the Sea* and its sequel through the Carolinas. But he planned to chop with a ruthless axe, and to broadcast his purpose as loudly as he could. No one else in the Union, certainly not Lincoln, had taken on this leading role as grim reaper. Sherman had come to understand both his enemy and the inner meaning of war as has no other American soldier. In a sense Lincoln could forgive because Sherman could destroy. Their relationship was both antagonistic and symbiotic. Sherman's scourging approach to the inner purposes of war in itself was as necessary to the Union victory as was Lincoln's belief that the final purpose of victory would be national conciliation. For the Confederates had to be defeated in devastating fashion, lest they sustain a longer term armed insurrection of the type which has characterized the histories of such peoples as the Irish and the tribes of the Balkans. In his ruthless marches through the Confederate heartland, at least as much ideologically as militarily, Sherman provided essential elements of the Confederate defeat. Ironically, Sherman's great campaigns served not only to win the war but to serve black freedom, even as he opposed that liberation, in which Lincoln so devoutly believed.

When Savannah fell to his army on December 22, Sherman wrote in triumph to Lincoln, "I beg to present you as a Christmas gift, the city of Savannah, with 150 heavy guns and plenty of ammunition, and also about 25,000 bales of cotton." Sherman was delighted when he received back Lincoln's message of gratitude. "When you were about leaving Atlanta for the Atlantic Coast, I was anxious if not fearful," Lincoln admitted, "but feeling that you were the better judge, and remembering that 'nothing risked, nothing gained,' I did not interfere. Now, the undertaking being a success, the honor is all yours; for I believe none of us went further than to acquiesce." This was vindication indeed, for the risk, for the self asserted independence, and for the conquest. If only after the fact, Lincoln appreciated Sherman's form of war-making. A loose command structure, with Sherman in effect running his own war, turned out to have had real benefits, even if it had been only spasmodically planned or directed from the top.[30]

Only after Lincoln's death did Sherman fully honor the fallen martyr, but even then he misinterpreted Lincoln's legacy in a willful effort to pursue another independent political policy, which he expressed in the separate peace treaty he forged with Confederate General Joseph Johnston a week after Lincoln's assassination. Sherman had available the model of Grant's terms, given to Lee on April 9, and approved by Lincoln, but he did not even attempt to replicate those terms. Although his thinking was not especially clear those two days of April 17 and 18, rather than imposing a warmly phrased but politically unconditional surrender as had Grant on Lee, Sherman made a fully political treaty with Johnston, ceding to the South breathtakingly generous military and political terms. In exchange for an agreement to "cease acts of war," Sherman granted a "general amnesty" to all remaining Confederate armies and not just Johnston's, and an amnesty as well to all Confederate civilians. The officers and their men were allowed to keep their arms, in order to take them to their several state arsenals for deposit. Sherman granted recognition by the Executive branch of the Union to all the existing Confederate legislatures, as reformed Union state governments, upon their taking an oath of allegiance to the Constitution of the United States, and he guaranteed to all southernors their "political rights and franchises, as well as all their rights of persons and property." Sherman was silent on the issue of the abolition of slavery, although he assumed that both he and Johnston knew the institution was dead. He had in effect readmitted the South to the Union without any political reconstruction, much less punishment or social reform. His treaty was rejected unanimously by the cabinet; it was far too conservative even for the extremely conservative new president, Andrew Johnston.[31]

As he approached these negotiations, Sherman had believed that in his desire to make a peace with generous political terms, he was reinforced by the wishes which the just assassinated president had expressed when they had met on March 27 and 28, off City Point Virginia, their first personal contact since since 1861. Sherman's later reconstructions of this conversation, particularly the one in his *Memoirs,* are self-serving and untrustworthy, especially when he concludes that Lincoln had "distinctly authorized me," to guarantee the civil rights and governments of the former Confederate states, "as soon as the rebel armies laid down their arms." In a private letter written in 1872, three years before the *Memoirs* version was published, Sherman's recollections read less like a 1875 political tract, and more like an account of the sort of rambling conversation Lincoln had held many times with other men. "He said that he contemplated no revenge, no harsh measures, but quite the contrary," Sherman wrote in this version. As for the actual words Lincoln used, Sherman had not thought at the time that this was such an important discussion, and therefore he admitted he had only hazy memories of them. "I cannot say that Mr Lincoln or anybody used this language at the time." Indeed his chief impression, which he stressed to his wife when he wrote to her soon after the event, was that Lincoln "was lavish in his good wishes" to Sherman. Lincoln was probably not concerned in spelling out detailed conditions of peace at this occasion. After all, Lincoln could not have anticipated the assassination and political brouhahas that were to follow. This meeting had seemed at the time more of a social visit than anything else. Therefore in reconstructing the event, Sherman fell back on the notion that he had caught Lincoln's drift. "I know I left his presence with the conviction that he had in mind, or that his cabinet had, some plan of settlement, ready for application the moment Lee & Johnson were defeated." Back in Washington, Lincoln was in fact engaged in an intense series of discussions over Reconstruction with Congress and his cabinet, and although he way playing his cards close to his vest, as was habitual with him, he was beginning to move toward accommodation with the Radicals, with whom he maintained constructive ties, and who had the clearest available policy. The Radicals were demanding an interim period of political change prior to a restoration of southern civil and political rights. Under the stresses of significant political negotiations, it would have been characteristic of Lincoln to have remained vague and general to his military leadership, if kindly in tone, for they were not really political players in Washington. It is highly unlikely that Lincoln gave Sherman any clear instructions about the political contents of a peace treaty. Nevertheless Sherman took into his discussions with Johnston a general sensibility that the martyred Lincoln would have favored the most generous possible terms.[32]

In the highly fluid political atmosphere that characterized the end of the war, Sherman was moving rapidly toward the most conservative possible position on reconstruction. If he had been the keenest destroyer in the war, his purpose had been unconditional Confederate surrender rather than political or social change. He had ignored Lincoln or crossed him during the war, and now he would quite cavalierly appropriate his version of Lincoln's memory to bolster his own reactionary political desires, as expressed in his separate peace treaty.

5

Grant, Lincoln, and Unconditional Surrender

John Y. Simon

GRANT AND LINCOLN

Bare-breasted Columbia, found so appealing by Victorian America, crowns the two greatest heroes of the Union with laurel wreaths. *(Lithograph by Kimmel and Forster, "The Preservers of our Union," 1865. Boritt Collection)*

ULYSSES S. GRANT FIRST visited Washington to pursue a claim for $1000 he lost during the Mexican War. As quartermaster, Grant was held liable for money stolen from the trunk of a brother officer in 1848. Cleared by a board of inquiry at the time, Grant remained responsible for repayment until relieved by act of Congress. In 1852, Lieutenant Grant, with a three-day leave, arrived in Washington to find the congressman from whom he expected help out of town for ten days and all offices closed for the funeral of Henry Clay. Frustrated in his business, Grant turned disappointed tourist, finding the city "small and scattering and the character of the buildings poor."[1] The debt hung over his head for another decade, until Congress rewarded the general who had captured Fort Donelson by relieving him of this unjust obligation.

Twelve years passed before Grant returned to Washington. On March 8, 1864, accompanied only by his teenage son, Grant registered at Willard's Hotel, where the clerk first assigned him an undesirable top floor room, and then fell all over himself when he looked at the registration book. Like almost everyone in Washington, the clerk knew that Grant had been promoted to lieutenant-general, a rank held previously only by George Washington by right and Winfield Scott by brevet, and had come to take command of all the armies of the United States. When Grant and his son Fred tried to eat dinner at the hotel, other guests gave him a hero's welcome. Grant fled before finishing his meal.

Worse followed that evening at the White House, where Grant attended a public reception thronged with people eager to see the new commander. After meeting Abraham Lincoln for the first time, Grant stood on a crimson sofa in the East Room shaking hands and receiving congratulations. He might have been trampled had he not ascended the sofa.

Later that evening, Lincoln took Grant aside to explain the ceremony scheduled for the following day. In the presence of the entire cabinet, Lincoln intended to present Grant with his commission as lieutenant-general and to make a brief speech, expecting Grant to reply. According to presidential secretary John Nicolay, Lincoln asked Grant to say something that would allay the jealousy of other generals and something to put him on "good terms" with the Army of the Potomac.[2] Grant's brief written remarks, three sentences to Lincoln's four, touched neither point directly, yet seemed entirely appropriate.

> I accept the commission with gratitude for the high honor confered. With the aid of the noble armies that have fought on so many fields for our common country, it will be my earnest endeavor not to disappoint your expectations. I feel the full weight of the responsibilities now devolving on me and know that if they are met it will be due to those armies, and above all to the favor of that Providence which leads both Nations and men.[3]

The next day, Grant visited the Army of the Potomac and its commander, Major-General George G. Meade, who had expected Grant to replace him with a veteran of the western armies. When Meade said that he would willingly relinquish command to another, perhaps Major-General William T. Sherman, and serve wherever placed, Grant promptly decided to retain Meade, a man who put service ahead of self. By the time he returned to Washington, Grant was ready to return west to confer with Sherman and to wind up affairs in Tennessee. To do so, he canceled dinner at the White House, an invitation he had already accepted, telling Lincoln that he had endured enough of "this show business."[4] Besides, he explained, "a dinner to me means a million dollars a day lost to the country."[5]

General-in-Chief Henry W. Halleck, now outranked by his former subordinate, had forced Grant's hand by requesting to be relieved. Higher rank required Grant to assume overall command. Grant resolved the problem by arranging Halleck's appointment as chief of staff, leaving him in Washington to coordinate orders, freeing Grant to establish headquarters wherever he wished. Halleck's new post separated strategic command from administration, a crucial innovation in modern warfare. Grant avoided the position of military adviser to Lincoln and Secretary of War Edwin M. Stanton, a role congenial to Halleck but intolerable to Grant, who intended to remain a commander rather than a courtier and to distance himself from politicians.

When he informed Sherman of his orders to report to Washington, Grant wrote that he would reject the promotion if required to stay there. As if unwilling to take Grant at his word, Sherman replied that Halleck was better able "to stand the buffets of Intrigue and Policy. . . . For Gods sake and for your Countrys sake come out of Washington."[6] But if Grant retained command in the West, as Sherman urged, he would supersede his most trusted lieutenant. Virtually without an alternative, Grant established headquarters with the Army of the Potomac, not displacing Meade but coordinating command from a point with the largest and most prominent of all armies, remaining close to Washington without becoming caught in its political eddies, and entrusting western command to Sherman.

Grant's visit to Washington lasted three days. During that time he conferred privately with Stanton and also with Halleck, dined with Secretary of State William H. Seward, and visited Meade. He spent little time with Lincoln and much of that in the presence of the cabinet. In their first private meeting, as Grant recalled it, Lincoln said that "he had never professed to be a military man or to know how campaigns should be conducted, and never wanted to interfere in them: but that procrastination on the part of commanders, and the pressure from the people at the North and Congress, *which was always with him,* forced him into issuing his series of 'Military Orders,' " some of which he conceded were mistaken. Lincoln added that "while armies were sitting down waiting for opportunities to turn up which might, perhaps, be more favorable from a strictly military point of view, the government was spending millions of dollars every day; that there was a limit to the sinews of war, and a time might be reached when the spirits and resources of the people would become exhausted." He simply wanted someone to "take the responsibility and act," something Grant assured him that he would do.[7] Lincoln said that he did not want to know what Grant proposed to do but brought out a plan of his own to use the Potomac as a base and advance between two streams that would protect the Union flanks. Grant realized, but did not mention to Lincoln, that these streams would also shield Lee. Thereafter, said Grant, he did not communicate his plans to Lincoln, Stanton, or Halleck.

Whether Lincoln abdicated his control of the war has since been disputed, but Grant thought that Lincoln did so. More likely, he gave this bulldog a longer leash than previous commanders, but a leash nonetheless. Facing the uncertainties of an election year, remembering that no incumbent had been re-elected since Andrew Jackson, Lincoln dared not risk responsibility for military setbacks. Grant's spring campaign opened with an exchange of formal letters with Lincoln. "The particulars of your plan I neither know, or seek to know," wrote Lincoln. "Should my success be less than I desire, and expect," Grant responded, "the least I can say is, the fault is not with you."[8]

Although Lincoln wrote that he did not know the "particulars" of Grant's plan, Lincoln had heard overall strategy explained in some detail. Grant intended to use all armies in a coordinated offensive. The Army of the Potomac would advance against Richmond from the north while the Army of the James under Major-General Benjamin F. Butler moved from Fort Monroe to threaten Richmond from the south. Major-General Franz Sigel would move up the Shenandoah Valley to counter any Confederate threat to Washington and, if successful, drive against Richmond from the west. Sherman's thrust into Georgia would be supported by an expedition to Mobile led by Major-General Edward R. S. Canby. Every army had orders to advance in early May. No longer could Confederates use interior lines to counter sporadic offensives. Every army would perform some role in the grand offensive. "Those not skinning can hold a leg," Lincoln observed to Grant, who repeated the same phrase, without attribution, in a letter to Sherman.[9] In devising this plan, Lincoln said privately, Grant implemented a strategy that Lincoln had always urged.

So Grant had a free hand not because Lincoln gave him one but because the president approved the strategy. Lincoln then led Grant to believe that he had more authority than he had in reality. By doing so, Lincoln enabled Grant to preserve the self-confidence that had brought victory in the West. At the same time, Lincoln separated himself from the carnage on the bloody road to Petersburg. He won unanimous renomination while bodies still lay on the field at Cold Harbor.

On May 4, Grant plunged across the Rapidan and led his armies into the Wilderness, suffering tremendous casualties. Moving on his left flank, he encountered Robert E. Lee again at Spotsylvania, at the North Anna, then at Cold Harbor, where the second assault proved singularly disastrous. Undaunted, Grant launched a brilliant flanking maneuver that bewildered Lee and took the Army of the Potomac south of the James River and Richmond to the thinly held Confederate lines at the vital rail center of Petersburg. Only then did Lincoln break six weeks of public silence about Grant's strategy and tactics. "I begin to see it. You will succeed—God bless you all."[10] Within a week, Lincoln visited Grant at his City Point headquarters.

Grant was the fourth man to command all the armies under Lincoln. The first, Winfield Scott, had filled the post since 1841. Despite age and infirmity he had served Lincoln well and retired on November 1, 1861, only because George B. McClellan insisted on holding the office himself. McClellan, in turn, lasted only four months, until March 11, 1862, when Lincoln, in effect, assumed the post himself on the eve of McClellan's ill-fated Peninsular campaign. McClellan's withdrawal from Richmond led to Halleck's arrival in Washington to restore shattered morale and to provide administrative control. Halleck's passive approach to strategy combined with his bureaucratic expertise suited Lincoln and Stanton. Eventually it suited Grant as well, for Halleck's role changed little when he moved from general-in-chief to chief of staff. Lincoln, however, had served as de facto general-in-chief for the past two years. He trusted some generals more than others, but trusted no general to see the war in the broad perspective afforded by the White House.

Grant had gone to Washington in March determined to return immediately to the West after accepting his commission. Once there, he changed his mind and decided to remain near Washington. Only the commanding general could "resist the pressure," he concluded, "to desist from his own plans and pursue others."[11] He proposed "to exercise actual command of all the armies, without any interference from the War Department . . ." He did not intend to become "McClellanized."[12] The fall of McClellan was so much on his mind that one week after leaving the White House he discussed with Sherman restoring McClellan to a major command.

Early in the war, Grant had sought a position on McClellan's staff. Grant spent two days waiting in McClellan's outer office in Cincinnati before taking the hint to look elsewhere for a command. "I knew McClellan," he recalled, "and had great confidence in him. I have, for that matter, never lost my respect for McClellan's character, nor my confidence in his loyalty and ability." He remained, said Grant, "one of the mysteries of the war." As a military professional, Grant remained sympathetic to McClellan because

> the test which was applied to him would be terrible to any man, being made a major-general at the beginning of the war. It has always seemed to me that the critics of McClellan do not consider this vast and cruel responsibility—the war, a new thing to all of us, the army new, everything to do from the outset, with a restless people and Congress. McClellan was a young man when this devolved upon him, and if he did not succeed, it was because the conditions of success were so trying. If McClellan had gone into the war as Sherman, Thomas, or Meade, had fought his way along and up, I have no reason to suppose that he would not have won as high a distinction as any of us. McClellan's main blunder was in allowing himself political sympathies, and in permitting himself to become the critic of the President, and in time his rival.[13]

It is impossible to recapture precisely what Grant meant by his apprehension of becoming "McClellanized." In part it expressed unwillingness to follow strategy dictated by the president and secretary of war, subordinating military to political considerations. Grant's new command entitled him to put Meade out to pasture, along with McClellan, a move Grant immediately rejected. In their initial conference, Lincoln referred to his war orders of two years earlier, admitting that they were, at least in part, mistaken. By implication, Lincoln indicated that he would not treat Grant as he had treated McClellan.

Later generations lauded Lincoln's wartime leadership and even military genius, but Grant accepted command in 1864 with a degree of wariness that gradually dissipated and ultimately was forgotten. Although Lincoln had appointed Grant a brigadier-general on July 31, 1861, even before the former captain, now a colonel, had encountered an armed enemy, he did so as a favor to Elihu B. Washburne, a political ally since 1843 and the ranking Republican in the House of Representatives. After a caucus of Illinois congressmen had decided how many of the thirty-four new brigadiers should come from Illinois, Washburne pushed Grant as the best-qualified soldier from his congressional district.

Grant's great victory at Fort Donelson, the North's first substantial victory, led to his promotion to major-general and a similar promotion for a principal subordinate, John A. McClernand, who had entered the war as Democratic congressman from Lincoln's home district. Caught unprepared for the Confederate assault at Shiloh in April, Grant rallied his forces for a second day of battle that more than redeemed the field. Lincoln withstood demands for Grant's removal but did not intercede when Halleck took personal command of his army and shunted aside Grant to frustrating inaction. Only Halleck's call to Washington as general-in-chief gave Grant another opportunity to command independently.

In September 1862, McClernand arrived in Washington to intrigue for command of an expedition to open the Mississippi River. To gain Lincoln's support he used the promise of new regiments of loyal Democrats and the threat of disloyalty in the Northwest if its normal commerce remained blocked. Lincoln succumbed but Halleck connived to send the new regiments to Grant at Memphis instead of holding them for McClernand's expedition. Grant sent Sherman on an ill-fated premature assault on Vicksburg designed to bag the Confederate citadel before McClernand arrived to claim command. Repulsed at Chickasaw Bayou, Sherman was superseded by McClernand. As quickly as possible, Grant arrived to supersede McClernand.

Grant ranked McClernand, but other corps commanders did not. Grant commanded a huge department; if he were called away to meet some threat elsewhere, McClernand would assume command of all forces before Vicksburg. At the suggestion of staff officer James Harrison Wilson, Grant wrote to Halleck proposing to merge the four departments in the Mississippi Valley under one senior commander, disclaiming any personal ambition for that post. As soon as Grant accepted that suggestion, Wilson began to scheme to bring the recently deposed McClellan west for a fresh start. Wilson recalled that Grant had no objection to serving under McClellan.[14] Lincoln, however, had no intention of giving McClellan further military opportunities, especially at Grant's expense.

In the early months of 1863, Grant endured the frustrations of flooded terrain that kept him from attacking Vicksburg and the presidential protection that kept him from sacking McClernand. Eventually Grant crossed the Mississippi River south of Vicksburg, launched a brilliant strike at Jackson, then drove the Confederates back to Vicksburg. Grant launched two assaults on Vicksburg before settling into a siege. Because of boastful misrepresentation by McClernand, the second assault proved more sanguinary than necessary. Constant friction between Grant and his ambitious subordinate ended with McClernand's removal two weeks before Vicksburg's surrender.

Lincoln had chosen McClernand as a fit partner for Rear-Admiral David D. Porter, since both condemned West Pointers as arrogant and uncooperative. Stanton and Halleck concurred; neither had confidence in Grant.[15] After his removal, McClernand bombarded Washington with statements in his defense, claimed credit for Grant's victories, and argued that his removal had originated in a clerical oversight. Grant sent John A. Rawlins, his adjutant and a prewar lawyer, to argue his case. Rawlins made a successful two-hour presentation to the cabinet. Rejoicing over the imminent opening of the Mississippi River, Lincoln no longer needed to heed cries of potential midwestern disaffection from a general who had demonstrated insubordination and incompetency.

So thoroughly had Lincoln changed that he wrote Grant a letter "of grateful acknowledgment for the almost inestimable service you have done." He had, he wrote, doubted Grant's strategy throughout and now wished "to make the personal acknowledgment that you were right, and I was wrong." This remarkable letter left Grant incapable of replying. Modesty halted his pen. Three weeks later Lincoln asked, "Did you receive a short letter from me, dated the 13th. of July?" "Your letter of the 13th of July was also duly received," Grant lamely responded.[16]

Presidential politics influenced the relationship between Lincoln and Grant. Democrats knew that the nomination of a successful general in 1864 represented their surest path to the White House. Like most professional soldiers of the prewar years, Grant had voted Democratic. The election of a Republican in 1856, Grant believed, would provoke secession and rebellion, so he had voted for James Buchanan. Besides, he explained, he knew John C. Frémont, the Republican candidate. Although uninvolved in politics, Grant had lost the post of county engineer of St. Louis in 1859 by a partisan vote of commissioners who believed him to be a Democrat. A recent move to Galena in 1860 cost Grant eligibility to vote; he favored Stephen A. Douglas. Grant's first military employment during the Civil War came from Republican governor Richard Yates of Illinois, who appointed him aide, mustering officer, then colonel, before Republican congressman Washburne arranged his appointment as brigadier-general. From that time onward, Grant's political affiliation remained unclear, a lack of clarity he found beneficial.

Just when Grant became a Republican has no easy answer. His wartime policy stressed military professionalism and adherence to federal mandates. Of all divisive issues, the most important centered on emancipation and its military counterpart, the use of black troops. Here Grant gave model adherence to administration policy, strengthened by his admiration for black troops of his own command who had withstood a Confederate attack on Milliken's Bend in June 1863.

After Grant's smashing victory at Chattanooga in November 1863, politicians of both parties appraised him as a potential 1864 presidential candidate. He firmly rebuffed Democratic overtures, the more urgent because the party, divided and leaderless since the death of Douglas in June 1861, desperately needed a military hero to oppose Lincoln.

Nearly three years of Civil War, with the end beyond sight and even the outcome in doubt, gave Lincoln concern about his renomination and re-election. Republican radicals massed against him eventually nominated Frémont as a protest candidate. Other Republican dissidents already looked to Grant. At the Republican convention in Baltimore in June, every vote went to Lincoln on the first ballot except twenty-two Missouri votes for Grant, later switched to make the nomination unanimous. By the summer of 1864, Grant could almost certainly have received the Democratic nomination for president or could have mounted a strong threat to Lincoln as a Republican.

Yet as late as 1868, Grant still did not want to be elected president. He intensely disliked newspaper attention, public speaking, ceremony, and every aspect of politics. With four children still in school, he craved the security of lifetime tenure in the peacetime army. Only Andrew Johnson through a relentless assault on congressional reconstruction and military authority could provoke Grant to action. Johnson first attempted to use Grant as a political pawn, then attacked him as faithless when he refused to comply. Even then, Grant referred to his enthusiastic supporters as "that awful Chicago Convention" and explained that he had been "forced into" accepting the nomination "in spite of myself."[17] Few have grasped his genuine reluctance to serve as president; little wonder that in 1864, Lincoln, a man supremely political, needed the most convincing evidence that Grant intended to remain in military command. Not until trusted political advisers like J. Russell Jones of Galena assured Lincoln that Grant harbored no presidential ambitions did Lincoln's concerns subside.

Congress created the rank of lieutenant-general for Grant but constitutional limitations prevented its naming him. Only Lincoln could nominate a lieutenant-general—and he could have promoted Halleck without changing the command structure. Even after assembling all available evidence about Grant's lack of presidential ambition, Lincoln remained wary.

An interpretation of the relationship between Lincoln and Grant that argues their immediate rapport falls far short of reality. Lincoln saw Grant as a potential political rival, as possibly another McClellan, that former military savior headed toward the Democratic nomination for president. Grant saw Lincoln as the commander-in-chief who had burdened McClellan with advice and orders, given his army to John Pope, and destroyed his military career. Nor could he have forgotten that Lincoln had given Sherman's command to McClernand, whose ambitions included toppling Grant. Before Grant came east, Lincoln needed abundant reason to believe that Grant posed no political threat. Grant needed equal assurance that Lincoln posed no military threat. Once the deal was struck, both needed reassurance that it had been made sincerely.

Lincoln twice visited Grant in Virginia during the summer of 1864. On June 20, Lincoln left Washington at 5:00 p.m. and arrived slightly seasick at City Point the next day about noon. After reviewing troops, he spent the evening sitting outside Grant's tent telling stories to the general and his staff. Accompanied by his son Tad, Lincoln had apparently come to visit, to inspect, to strengthen a personal relationship, but hardly as commander-in-chief. Lincoln filled the following day with inspections and reviews before leaving for Washington. During the visit, the two had spent little time together; what they discussed and its effects on strategy remain unknown. Perhaps the meeting takes on significance largely through its lack of significance: Grant and Lincoln reassured each other of good intentions by maintaining separate spheres of authority.

Grant's brilliant movement across the James River to Petersburg had been followed by federal folly, incompetence, and timidity as thinly held Confederate lines stalled an assault of tenfold more troops. By the time Lincoln arrived on June 21, Grant had lost the momentum for a smashing victory and faced Lee's veterans in the Petersburg trenches. Furthermore, Jubal Early's corps, sent to stop David Hunter's threat to Lynchburg, had driven federal forces into West Virginia and could look down the Shenandoah Valley toward Washington. Gradually Grant grasped the dismal implications of those missed opportunities and settled into a siege. Meanwhile, he underestimated Early's threat, detaching troops only after Early reached Harper's Ferry. Lew Wallace, defeated at Monocacy, Maryland, on July 9, may have saved the capital by delaying Early's advance on Washington for a single day. By the time

Confederate troops reached the outskirts of Washington, enough of Grant's tardily sent troops had reached the forts to discourage more than a skirmish. Yet Early's corps remained intact, threatening numerous important transportation and supply centers while weakening northern morale. To coordinate the pursuit of Early, Grant first had suggested William B. Franklin. On July 25, he recommended placing Meade in charge, for reasons he did "not care to commit to paper."[18]

Smoldering animosities blazed in the Army of the Potomac, fanned by William F. Smith, the commander who deserved blame for the June 15 failure at Petersburg but who blamed everyone else. Relentless in charges against both Meade and Butler, Smith implicated Grant himself in failure. Both Meade and Butler reacted vigorously to criticism, Meade by loudly losing his temper, Butler by quietly plotting against his foes.

To break the Petersburg siege, coal miners of the 48th Pennsylvania dug a tunnel beneath Confederate lines, filled the end with explosives, and detonated a devastating blast. Union troops failed to exploit the opening and were repulsed with heavy losses, a battle Grant called the "saddest affair I have witnessed in this war. Such opportunity for carrying fortifications I have never seen and do not expect again to have."[19]

The day after the battle of the Crater, with Early still a threat to northern cities, Lincoln arrived at Fort Monroe for a five-hour conference with Grant. So little is known of their discussion that the noted historian T. Harry Williams wondered if they had actually met; yet they did meet. This time, however, Lincoln came without Tad, Grant without his staff. Lincoln neither reviewed troops nor regaled an appreciative audience with stories. Grant omitted all mention of this meeting in his *Memoirs*.

Grant wrote his *Memoirs* with rare skill, modesty, balance, and candor. Desperately ill, he intended to leave a legacy for his impoverished family and a testament to his countrymen. Nonetheless, he summarized in six pages his six years in St. Louis from his resignation in 1854 until his move to Galena in 1860, years of struggle and hardship. These few pages contain more information about the politics of the period than of Grant's life. In an otherwise detailed Civil War narrative, he did not allude to General Orders Number 11, which he issued in December 1862. These orders, ostensibly designed to regulate trade, expelled all Jews from the Department of the Tennessee. Responding to vigorous protests, Lincoln revoked this outrageous act of bigotry. During the 1868 presidential campaign, Grant grudgingly admitted his error—the only Democratic charge against him that he answered—but chose not to revisit this painful incident in his *Memoirs*.

Grant's silence about the Fort Monroe meeting speaks eloquently of the pain of Lincoln's rebuke. Grant had been slow and ineffective in meeting Early's threat and had not yet coordinated a unified pursuit. After leading his army on the bloody road to Petersburg, he had not harvested the reward of his strategy. He had been vacillating and tentative in restructuring his command team and had ignored political realities tied to generalship. Although nothing is known of the actual content of the conference, its consequences were clear enough. If Grant had entered the Wilderness with any misconception that Lincoln had delegated management of the war, Grant received the necessary correction. Lincoln took control.

On the back of Grant's telegram arranging the meeting, Lincoln jotted a few words, presumably the agenda for the conference. "Meade & Franklin" represented Grant's suggestions for unified command to defend the capital; "Md. & Penna." indicated how broad that command might be. Between them Lincoln wrote the single word "McClellan."[20] Restoring McClellan to command had several advantages: enlisting the talents of the general who had twice before saved the capital, spurring northern enlistments, and depriving Democrats of their most attractive potential presidential nominee. With Grant holding rank as lieutenant-general and general-in-chief, McClellan would take the field as a subordinate. The Blair family had begun to negotiate with McClellan. The plan foundered because McClellan refused to take himself out of contention for the nomination or to ask Lincoln for command. While McClellan continued to sulk, however, Lincoln and Grant may well have discussed McClellan's military future. Such executive speculation reflected diminished faith in Grant.

As in 1862, when McClellan was as close to Richmond as Grant was two years later, the result of Early's raid might have been a detachment of troops from offensive to defensive deployment. Before Lincoln arrived, Grant had ordered a division of cavalry to Washington. After the conference, he sent Major-General Philip H. Sheridan to take command of all troops in the area and "to put himself south of the enemy and follow him to the death."[21] When Lincoln read the telegram, he responded that such action would "neither be done nor attempted unless you watch it every day, and hour, and force it."[22] Grant hurried to Washington to organize the pursuit of Early, something he disliked doing since Butler was senior in command whenever Grant was absent.

In effect Lincoln had told Grant where to go and what to do, and had at least hinted what might happen if he failed the assignment. With twice as many troops as Lee, Grant had settled into a protracted siege and had allowed himself to become "McClellanized." Furthermore, Lincoln analyzed the military situation more clearly than did Grant. Lee had pinned Grant at Petersburg effectively enough to detach Early's corps for a major northern campaign, with enough men to sweep down the Shenandoah Valley and to approach the defenses of Washington. If Lincoln's intervention had been unseemly, it was timely and necessary. Further, he had withheld intervention until the case for defending Washington was airtight. Even Grant recognized the necessity for an offensive in the Shenandoah Valley.

He also recognized that the president expected greater care in assigning generals. Of the commanders expected to lead major armies in the 1864 spring offensive, three owed their positions to political factors: Nathaniel P. Banks, Butler, and Sigel. None would survive Grant's drive for military professionalism, none would be easy to shelve. Banks's failure in the ill-conceived and badly executed Red River expedition led Grant to telegraph to Halleck that he had "been satisfied for the last nine months that to keep General Banks in command was to neutralize a large force and to support it most expensively." "Although I do not insist on it," he wrote, he wanted Major-General Joseph J. Reynolds to replace Banks.[23] Halleck replied that Lincoln wanted to delay to assess the results of the expedition; furthermore, William

B. Franklin would be the senior officer in the Department of the Gulf when Banks fell. If Banks could be ordered back to New Orleans, Grant conceded, placing the troops under the "senior officer" would suit him. Franklin had been "mixed up with misfortune," but was "much better than Gen Banks."[24] In response, Halleck explored the complexities.

> I submitted your telegram of 10.30 A. M. to the secty of War, who was of the opinion that, before asking the President for an order, I should obtain your views in regard to the extent of the proposed Division, the officer to command it, &c., and that I should write to you confidentially on the subject. Do you propose to include Pope's Curtis' & Rosecrans' commands, or only the present Depts of the Gulf and of Arkansas with the Indian Territory? Is it proposed to give Banks the command of the Division, or to leave him in the subordinate position of his present Dept, or to remove him entirely? In either case, the order must be definite. If Banks is superceded, Franklin will be the ranking officer in the field, and Rosecrans, Curtis or McClernand, in the Division. You have also heretofore spoken of Steele and Reynolds in connexion with this command. I think the President will consent to the order, if you insist upon Genl Banks removal as a military necessity, but he will do so very reluctantly, as it would give offence to many of his friends, & would probably be opposed by a portion of his cabinet. Moreover, what could be done with Banks? He has many political friends who would demand for him a command equal to the one he now has. The result would probably be the same as in the cases of Rosecrans, Curtis, Sigel, Butler & Lew. Wallace. Before submitting the matter to the President, the Secty of War wishes to have in definite form precisely the order you wish issued.[25]

Grant countered with a suggestion that Halleck himself take command of the "trans-Mississippi Division," at least temporarily turning over his Washington office to Canby. Halleck responded that he would willingly serve "any where and every where" but Canby was quite ill.[26] The president, fully aware of Grant's suggestions, had said nothing. When Grant firmly insisted on Banks's removal, he learned that Canby had already taken command. When Halleck contemplated leaving Washington for service in the field, Canby's health immediately improved. Before Canby went to New Orleans, the extent of Banks's failure was fully known. His removal, however, required more than an expression of Grant's wishes; he had to take full responsibility. His first telegram was unequivocal about the removal of Banks, however, and the president required emphatic repetition of this request.

Appointed to command in the Shenandoah to please a German constituency, Sigel soon demonstrated an inability even to hold a leg. When Early appeared, Sigel disappeared. Grant urged Halleck to relieve Sigel of command, "at least until present troubles are over."[27] His replacement was another matter. Grant suggested Major-General Edward O. C. Ord, but through lack of response from Washington, Hunter took command, although Stanton considered him "far more incompetent than even Sigel."[28] Grant's recommendation that Franklin or Meade assume command brought Lincoln to the Fort Monroe conference and resulted in Sheridan's departure for the Shenandoah Valley.

From the time Grant took command he recognized the problem posed by Butler. A prominent prewar Democrat who had rushed to war and received the reward of the rank of major-general of volunteers so early that he claimed to rank all other active generals except Grant, Butler had sedulously preserved political alliances while creating havoc throughout the army. Recognizing the necessity of maintaining Butler in the field in an election year, Grant had sent him two corps commanders for the Army of the James, William F. Smith and Quincy A. Gillmore, believed to represent the best military professionalism. During the campaign of May 1864, both Smith and Gillmore had failed Butler with resulting neutralization of 30,000 troops "bottled up" at Bermuda Hundred. Once Grant crossed the James River, the Army of the Potomac and the Army of the James united, making Butler senior commander under Grant, and commander by regulation whenever Grant was absent, as he rarely was for that very reason.

In this matter, Grant had strong allies in Washington. Both Stanton and Halleck so detested Butler that the former sent Assistant Secretary of War Charles A. Dana to City Point to suggest Butler's transfer. In a letter to Halleck, Grant danced daintily around the issue.

> Whilst I have no diff[i]culty with Gen. Butler, finding him always cle[ar] in his conception of orders, and prompt to obey, yet there is a want of knowledge how to execute, and particularly a prejudice against him, as a commander, that operates against his usefulness. . . . As an administrative officer Gen. Butler has no superior. In taking charge of a Dept.mt where there are no great battles to be fought, but a dissatisfied element to controll no one could manage it better than he. If a command could be cut out such as Mr Dana proposed, namely Ky. Ill. & I[ndian]a. or if the Depts. of the Mo. Kansas and the states of Ill. & Ia. could be merged together and

Gen. Butler put over it I believe the good of the service would be subserved.[29]

Halleck answered with unusual frankness.

I will, as you propose, await further advices from you before I submit the matter officially to the Secty of War and the President. It was foreseen from the first that you would eventually find it necessary to relieve Genl. B. on account of his total unfitness to command in the field, and his generally quarrelsome character. *What shall be done with him,* has therefore already been, as I am informed; a matter of consultation. To send him to Kentucky would probably cause an insurrection in that state, and an immediate call for large reenforcements. Moreover, he would probably greatly embarras Sherman, ~~or~~ if he did not attempt to supersede him, by using against him all his talent at political intrigue and his facilities for newspaper abuse. If you send him to Missouri, nearly the same thing will occur there. Although it might not be objectionable to have a free fight between him and Rosecrans, the Government would be seriously embarrassed by the local difficulties, and calls for reinforcements likely to follow. Inveterate as is Rosecran's habit of continually calling for more troops, Butler differs only in demanding instead of *calling*. As things now stand in the west, I think we can keep the peace; but if Butler be thrown in as a disturbing element, I anticipate very serious results. Why not leave Genl Butler in the local command of his Dept, including N. C. Norfolk, Fort Monroe, Yorktown, &c, and make a new army corps of the part of the 18th under Smith? This would leave B. under your immediate control, and at the same time would relieve you of his presence in the field. Moreover, it would save the necessity of organizing a new Dept. If he must be relieved entirely, I

> think it would be best to make a new Dept. for him in New England. I make these remarks merely as suggestions. Whatever you may finally determine on, I will try to have done. As Genl. B. claims to rank me, I shall give him no orders wherever he may go, without the special direction of yourself or the Secty of War.[30]

Grant eagerly grasped Halleck's plan to shelve Butler, and Lincoln approved the orders; but Grant ultimately backed down for reasons that have remained controversial. Smith claimed that Butler blackmailed Grant after catching him drinking; others have argued that Grant realized that Smith's unrelenting criticism of other generals, including Meade and perhaps Grant himself, led to recognition that Smith was a greater hazard than even Butler. Smith left, Butler stayed. Lincoln tried to avoid involvement; if Butler left, Grant would have to send him away.

In late June, Grant responded with disgust to a message that Major-General William S. Rosecrans and Major-General Samuel R. Curtis had called for reinforcements in Missouri. "I am satisfied you would hear the same call if they were stationed in Maine. The fact is the two Depts. should be merged into one and some officer, who does not govern so largely through a secret police system as Rosecrans does, put in command. I do think the best interests of the service demands that Rosecrans should be removed and some one else placed in that command. It makes but little difference who you assign it would be an improvement."[31] Nonetheless, it did make a difference who inherited Rosecrans's department. Furthermore, Rosecrans, a radical favorite, was the best-known Catholic general, hardly a factor to ignore in an election year.

Following the Fort Monroe conference, Grant no longer proposed to shuffle generals like a pack of cards. Rosecrans continued to mismanage Missouri until after the election. Meade remained in command of the Army of the Potomac, and Butler lost the Army of the James only after demonstrating clear incompetence at Fort Fisher—and after the election.

In August, despairing of both political and military victory, Lincoln prepared a memorandum for his cabinet to sign, sight unseen. "This morning, as for some days past, it seems exceedingly probable that this Administration will not be re-elected. Then it will be my duty to so co-operate with the President elect, as to save the Union between the election and the inauguration; as he will have secured his election on such ground that he can not possibly save it afterwards."[32] Lincoln intended, he explained later, to call upon McClellan after his election victory to raise troops for one last desperate effort for victory. By implication at least, Lincoln had decided that Grant could not win the war; only McClellan could defeat the Confederates.

Within a few days, however, Lincoln learned that Sherman could finish the job. The fall of Atlanta came precisely in time to counter Democratic claims that the war was a failure. Sherman then planned a march to the sea, a move so daring that even Grant hesitated before giving approval. Rawlins vigorously opposed the expedition and, as Grant recalled, persuaded authorities in Washington to delay Grant's approval. Stanton telegraphed on October 12 that the president expressed "much solicitude" about the plan and hoped that "it will be maturely considered."[33] Grant had already wired his approval. Sherman's celebrated

Christmas gift of Savannah to Lincoln was equally Grant's triumph.

On February 16, 1862, Grant had demanded the "unconditional and immediate surrender" of Fort Donelson. Confederate Brigadier-General Simon B. Buckner accepted these "ungenerous and unchivalrous terms" which roused considerable enthusiasm in the North and gave Grant an enduring nickname since unconditional surrender matched the initials with which he signed official correspondence. Given the circumstances of a besieged fort from which two ranking commanders had fled with a portion of the garrison, Grant did not need to dicker with Buckner, to whom he showed considerable courtesy after the surrender. Under similar conditions, McClellan himself might have demanded unconditional surrender, though he had less aptitude for phraseology.

Capturing another Confederate army at Vicksburg, Grant again demanded unconditional surrender but in fact negotiated with Major-General John C. Pemberton. Grant's decision to parole rather than imprison the garrison received some criticism despite logistical factors sustaining his judgment. Nonetheless, Grant came to Washington with his reputation intact for pushing for unconditional surrender.

By the end of 1864, after Lincoln's re-election, Sherman's capture of Savannah, and Major-General George H. Thomas's smashing victory at Nashville, the end of the war loomed. Efforts to negotiate peace during 1864, sometimes bizarre and always doomed, at least uncovered the two basic Confederate demands: independence and slavery. When hope for independence waned, that for slavery persisted.

Confederates thought that northerners might sacrifice emancipation for peace and reunion. Under such circumstances, Lincoln stood for unconditional surrender and wondered whether Grant would stand firm.

During the summer of 1864, Grant had realized the importance of Lincoln's re-election. In a private letter to Washburne, clearly intended for public use, Grant unmistakably endorsed Lincoln without mentioning his name or his party.

> I state to all Citizens who visit me that all we want now to insure an early restoration of the Union is a determined unity of sentiment North. The rebels have now in their ranks their last man. The little boys and old men are guarding prisoners, guarding rail-road bridges and forming a good part of their garrisons for intrenched positions. A man lost by them can not be replaced. They have robbed the cradle and the grave equally to get their present force. Besides what they lose in frequent skirmishes and battles they are now loosing from desertions and other causes at least one regiment per day. With this drain upon them the end is visible if we will but be true to ourselves. Their only hope now is in a divided North. This might give them reinforcements from Tenn. Ky. Maryland and Mo. whilst it would weaken us. With the draft quietly enforced the enemy would become dispondent and would make but little resistence. I have no doubt but the enemy are exceedingly anxious to hold out until after the Presidential election. They have many hopes from its effects. They hope a counter revolution. They hope the election of the peace candidate. In fact, like McCawber, the hope *something* to turn up. Our peace friends, if they expect peace from separation, are much mistaken. It would be but the begining of war with

> thousands of Northern men joining the South because of our disgrace allowing separatio[n.] To have peace ~~the~~ "on any terms" the South would demand the restoration of their slaves already freed. They would demand indemnity for losses sustained, and they would demand a treaty which would make the North slave hunters for the South. They would demand pay or the restoration of every slave escaping to the North.[34]

This letter received wide circulation as Republican campaign material, including an appearance in a pamphlet entitled *Democratic Statesmen and Generals to the Loyal Sons of the Union*. Grant's status as a Democrat added strength to the Union ticket of 1864 on which Lincoln ran with Johnson, an unrepentant Democrat. In September, Washburne asked for permission to circulate the letter that Grant had written to Lincoln on the eve of the campaign. Grant replied, "I have no objection to the President using any thing I have ever written to him as he sees fit—I think however for him to attempt to answer all the charges the opposition will bring against him will be like setting a maiden to work to prove her chastity—"[35] Grant also helped Lincoln by facilitating voting in the field. He congratulated Lincoln indirectly on re-election, emphasizing his relief that there had been no disorder at the polls.

Even after the election, Lincoln had no reason to forget that Grant was a prewar Democrat and slaveholder. As a professional soldier, Grant shared the conservatism of the peacetime army. Fifteen years in the army starting at West Point had provided him with close friends now commanding Rebel forces. Marriage into a family of border state slaveholders strengthened his southern ties, including those to his wife's cousin James Longstreet, previously Grant's old army friend. Grant had demanded unconditional surrender of armies when he had no practical alternative. How would he respond to peace initiatives and offers of negotiation?

The test came at the end of January 1865, when Confederate emissaries arrived at City Point requesting permission to visit Washington to confer with Lincoln about the "existing War," and "upon what terms it may be terminated." The impressive delegation consisted of Confederate Vice President Alexander H. Stephens, once a Georgia Whig congressman admired by fellow Whig congressman Lincoln; Confederate Assistant Secretary of War John A. Campbell, former justice of the U.S. Supreme Court; and Robert M. T. Hunter, prewar U.S. senator from Virginia, late Confederate secretary of state. Lincoln sent Major Thomas T. Eckert to pose conditions: unless the commissioners discussed a "common country," no conference could take place. Grant, who had served dinner to the commissioners, now realized that he had blundered through cordiality. To underscore the point, Lincoln told Grant to allow "nothing which is transpiring, change, hinder, or delay, your Military movements, or plans."[36]

In a telegram to Stanton, Grant asserted that Stephens and Hunter had shown him "that their intentions were good and their desire sincere to restore peace and Union."[37] He regretted that Lincoln did not plan to meet at least these two. Grant forced Lincoln's hand. "Induced by a despatch of Gen. Grant," he wired Seward, "I Join you at Fort-Monroe so soon as I can come."[38] The conference itself, to which Grant was not invited, proved entirely unproductive. Lincoln demanded reunion and emancipation;

Jefferson Davis had empowered his emissaries to concede neither but only to urge a hare-brained scheme to unite in an expedition to drive the French from Mexico. Although Lincoln suggested that he might favor compensation to slaveholders after rebellion ceased, Confederates characterized this policy as first demanding "unconditional submission." Indeed, the Confederates were correct. Lincoln gained nothing by conferring and would not have gone without Grant's urging, a point he emphasized in reporting to Congress.

Later in February, Grant was drawn into one of the silliest episodes of the war. Confederate commissioners at City Point had been favorably impressed by Grant's friendliness and that of Mrs. Grant, who hoped that the Confederates might be inveigled into releasing her brother John Dent, held prisoner in the South despite his ardent Rebel sympathies. Grant refused to exchange a soldier for his civilian brother-in-law, especially since Dent had foolishly thought that outspoken support of the Confederacy gave him license to travel freely in the South. Remembering Mrs. Grant's cordiality, and perhaps forgetting its cause, Julia's cousin Longstreet proposed to Ord that Mrs. Longstreet and Mrs. Grant exchange social visits as a first step toward conversations between officers of both sides that might end with Grant and Lee suspending hostilities to negotiate. Grant dismissed the proposal as "simply absurd."[39]

Conversations between Ord and Longstreet on political prisoners veered toward peace negotiations. Longstreet asserted, Ord recorded, that Lee believed the southern cause to be hopeless while Davis insisted on continuing the war. Ord suggested that Lee threaten to resign so as to force

Davis to negotiate for compensation for lost slaves "and an immediate share in the Gov't."[40] When Lee requested a meeting with Grant to discuss matters further, Grant forwarded the message to Washington and received an unequivocal reply from Stanton. "The President directs me to say to you that he wishes you to have no conference with Gen Lee unless it be for the capitulation of Lees army, or on solely minor and purely military matters He instructs me to say that you are not to decide, discuss, or confer upon any political question: such questions the President holds in his own hands; and will submit them to no military conferences or conventions—mean time you are to press to the utmost, your military advantages." In his own name, Stanton elaborated.

> I send you a telegram written by the President himself in answer to yours of this evening which I have signed by his order. I will add that General Ords conduct in holding intercourse with General Longstreet upon political questions not committed to his charge is not approved. The same thing was done in one instance by Major Key when the army was commanded by General McClellan and he was sent to meet Howell Cobb on the subject of exchanges. and it was in that instance as in this disapproved. You will please in future instruct officers appointed to meet rebel officers to confine themselves to the matters specially committed to them.[41]

Again, Lincoln rather than Grant pushed toward the goal of unconditional surrender.

In January, Grant had discussed with Lincoln the awkward status of the president's son Robert, a student at Harvard College and Law School during a war to which his father had sent so many other sons. His father wrote that Robert "wishes to see something of the war before it ends. I do not wish to put him in the ranks, nor yet to give him a commission, to which those who have already served long, are better entitled, and better qualified to hold. Could he, without embarrassment to you, or detriment to the service, go into your Military family with some nominal rank, I, and not the public, furnishing his necessary means? If no, say so without the least hesitation, because I am as anxious, and as deeply interested, that you shall not be encumbered as you can be yourself."[42] Grant responded that he would "be most happy to have him in my Military family . . ." Rank, he continued, would be "immaterial but I would suggest that of Capt. as I have three Staff officers now, of conciderable service, in no higher grade. Indeed I have one officer with only the rank of Lieut. who has been in the service from the begining of the war. This however will make no difference and I would still say give the rank of Capt.—"[43] The next day, Robert wrote that he needed to return to Cambridge and then wanted to attend his father's second inauguration, so he asked Grant's "kind indulgence" before reporting.[44] Before he did so, his father had formally nominated him for a commission as captain. If Robert's princely status and aristocratic airs annoyed Grant, he gave no indication of displeasure. Robert's staff appointment, so unwarranted a favor, cemented through an ignoble transaction a personal relationship between Grant and Lincoln previously nonexistent.

Robert's prolonged civilian life and military sinecure probably owed much to his mother's increasing emotional deterioration. She had never recovered from her son Willie's death in February 1862, and her behavior grew steadily more irrational and intolerable. Perhaps unaware of the extent of the problem, Grant invited Lincoln to visit in March, thinking that he would want to see Robert but would not do so without a formal invitation. Besides, Robert had no staff role greater than escorting his parents.

Accompanied by Mary and Tad, Lincoln arrived at City Point on March 24 for a visit that lasted a fortnight. Jealous of Grant's growing fame, Mary almost immediately insulted Julia. On a carriage ride to inspect Ord's division, Mary launched into a hysterical outburst that embarrassed everybody. For the remainder of the visit, Julia avoided Mary, who left for Washington on April 1, only to return on April 6.

Lincoln, Grant, Sherman, and Admiral Porter met on March 28 on the president's boat, the *River Queen*. With the end of the war finally in sight, the chieftains met to contemplate the future. What took place remains obscure. In his *Memoirs,* Grant omitted all mention of events during that part of the president's visit when Mary accompanied him. Sherman and Porter remembered that Lincoln was ready to recognize existing southern governments whenever the war ended, with "liberal views" toward Rebels, and "peace on almost any terms."[45] Both Sherman and Porter, however, sought to justify Sherman's overly generous terms to the Confederacy's General Joseph E. Johnston, given the day Lincoln died and quickly disavowed by Johnson and Stanton. Sherman had negotiated the surrender of all Confederate armies, had provided that their weapons could be taken to their state arsenals, and had stipulated that the United States would recognize the authority of these state governments and the political and property rights of their citizens. Had Sherman's terms received approval, slavery might well have survived the war when former Confederate states refused to ratify the Thirteenth Amendment.

If Sherman had been overwhelmed by the congratulatory and harmonious mood in the cabin of the *River Queen,* Grant remembered the harsh tone of the rebuke received when he proposed a military convention with Lee. He also remembered the two points made to the Confederate commissioners in February: that "the Union should be preserved" and that "slavery should be abolished."[46] These were preconditions for negotiation rather than negotiable. Porter, who unconvincingly claimed to remember the conversation on the *River Queen* in great detail, reported that Grant sat quietly smoking throughout, speaking only once, to ask about Sherman's destruction of railroads.

For that matter, Grant was absorbed in planning his final offensive, launched at the end of that month. When Petersburg fell on April 2, Grant invited Lincoln to meet him inside the town. Accompanied by Robert and Tad, Lincoln arrived the next day and strode to a house commandeered by Grant. As the president and Grant sat quietly on the porch the yard filled with former slaves, staring worshipfully at their liberators. Again the two found little to say. Grant remembered only that Lincoln had the right to ask about campaign plans, something he refrained from doing, and Grant did not volunteer information. Yet Lincoln never took his eyes off the war. As he left for Washington, he telegraphed to Grant. "Gen. Sheridan says 'If the thing is pressed I think that Lee will surrender.' Let the *thing* be pressed."[47]

Lincoln left City Point on April 8 and arrived in Washington the next evening to find streets filled with people celebrating Lee's surrender at Appomattox. After the war, artist William Page, once hailed as "the American Titian," planned his masterpiece, a giant canvas of Grant entering Richmond in triumph. A request that Grant pose brought the response that Grant had never entered Richmond. In April 1865 he had hurried off to Washington to finish off the war. On April 13, he received a note from Mary Lincoln inviting him "to drive around with us to see the illumination."[48] During the drive, cheers for Grant so provoked Mary to abusive language that Grant decided against accompanying the Lincolns to the theater the following evening, deciding instead to visit his children in New Jersey. Aboard the train, the Grants learned of Lincoln's assassination.

Had Lincoln lived, Sherman insisted, the surrender terms to Johnston would have received presidential approval. Grant knew better. He made excuses for Sherman and embraced Sherman's quarrels with Stanton and Halleck without deviating from the central principle that Sherman had no executive warrant for terms beyond those given to Lee at Appomattox. Those terms, however tempered by graciousness and compassion, amounted to unconditional surrender. Lincoln had insisted upon that.

When all their meetings are summed up, Lincoln and Grant had spent little time together. Because of the problem of Mrs. Lincoln, Grant (and others) sometimes avoided the Lincolns. Grant remembered Lincoln's ready wit and storytelling ability better than some frustrations in implementing military policy. To interpret their relationship in terms of an immediate understanding that created an instantaneous partnership is to miss the point of Lincoln's leadership. He held the reins and taught Grant what was permitted and what was forbidden.

Lincoln's tragic death led to his enshrinement in the memories of those who had known him, even those who had clashed with him. A bitter quarrel with Johnson enhanced Grant's favorable opinion of Lincoln. Grant's presidency as Lincoln's Republican successor provided additional reason for admiration. Finally, Grant's post-presidential role as party leader elicited further praise of Lincoln. In writing and speaking in later years, Grant succumbed to the sentimentality of the age. He created the impression that mutual harmony and respect had existed from their first meeting and persisted until Lincoln's death.

In reality, however, Grant and Lincoln forged an effective partnership in a turmoil of clashing authority. Amidst the confusion of war, Lincoln redefined the concept of commander-in-chief, Stanton transformed the concept of secretary of war, and Halleck created the concept of chief of staff. Charged with vast responsibilities, General-in-Chief Grant had to act vigorously within the military sphere, tread softly in the political sphere, and understand as well the politics of command. Under Lincoln's guidance, sometimes oblique, sometimes imperious, Grant succeeded.

Notes

Introduction

1. Robert E. Lee to Mrs. Mary Custis Lee, July 12, 1863, in Clifford Dowdey, ed., Louis H. Manarin, assoc. ed., *The Wartime Papers of R. E. Lee* (Boston: Little, Brown, 1961), 547.

2. Diary of John Hay, July 19, 1863, Hay Papers, Brown University. The printed version of this diary contains significant inaccuracies for the period of the Gettysburg campaign: Tyler Dennett, ed., *Lincoln and the Civil War in the Diaries and Letters of John Hay* (New York: Dodd, Mead, 1939).

3. *Boston Daily Advertiser,* July 20, 1865.

4. Roy P. Basler, ed., Marion Dolores Pratt and Lloyd A. Dunlap, assist. eds., *The Collected Works of Abraham Lincoln,* 9 vols. (New Brunswick, 1953–63), 6:78–79.

5. Ibid., 328.

6. George Gordon Meade to Mrs. Margaretta Sergeant Meade, Dec. 2, 1846, Meade Papers, Historical Society of Pennsylvania.

7. William T. Sherman to William M. McPherson, [c. Sept., 1864], Sherman Papers, Huntington Library, San Marino, California.

One: Lincoln and McClellan

1. Brian C. Pohanka to author, Nov. 19, 1985.

2. *Philadelphia Ledger,* Oct. 30, 1885.

3. A. K. McClure, *Abraham Lincoln and Men of War-Times* (Philadelphia: Times Publishing, 1892), 192.

4. *McClellan's Own Story* draft; John M. Douglas to McClellan, Jan. 26, 1861, McClellan Papers, Library of Congress.

5. McClellan to Scott, Apr. 27, 1861, Stephen W. Sears, ed., *The Civil War Papers of George B. McClellan: Selected Correspondence, 1860–1865* (New York: Ticknor & Fields, 1989), 12–13; Scott to McClellan, May 3, 21, 1861, *The War of the Rebellion: A Compilation of the Official Records of the Union and Confederate Armies,* 127 vols., index and atlas (Washington: GPO, 1880–1901), ser. 1, vol. 51, pt. 1, pp. 369–70, 386–87. Hereafter *OR*.

6. McClellan proclamation, May 26, 1861; McClellan to Jacob Beyers, July 14, 1861; McClellan to Lincoln, May 30, 1861, *McClellan Papers,* 26, 53–54, 28.

7. McClellan to Mrs. McClellan, July 27, 1861, *McClellan Papers,* 70; Benjamin P. Thomas, *Abraham Lincoln: A Biography* (New York: Knopf, 1952), 458, 460.

8. William Howard Russell, *My Diary North and South* (1863; reprint New York: Harper, 1954), 256–57; Samuel P. Heintzelman diary, Nov. 11, 1861, Heintzelman Papers, Library of Congress; Tyler Dennett, ed., *Lincoln and the Civil War in the Diaries and Letters of John Hay* (New York: Dodd, Mead, 1939), 35.

9. McClellan to Mrs. McClellan, Oct. 31, Oct. 16, Nov. 17, 1861, *McClellan Papers,* 113, 107, 135–36.

10. *McClellan's Own Story* draft, McClellan Papers; McClellan to Mrs. McClellan, Aug. 29, 1862, McClellan Papers, 419; Thomas, *Lincoln,* 458.

11. McClellan to Scott, Aug. 8, 1861; to Cameron, Sept. 13, 1861, to Stanton, June 25, 1862, *McClellan Papers,* 79–80, 100, 309; Edwin C. Fishel, "Pinkerton and McClellan: Who Deceived Whom?," *Civil War History,* 34 (1988): 116–17, 128; McClellan to Mary Ellen Marcy, c. 1859, McClellan Papers.

12. Scott to Cameron, Aug. 9, 1861, *OR,* ser. 1, vol. 11, pt. 3, p. 4; *Harper's Weekly,* Aug. 24, 1861; McClellan to Cameron, Sept. 13, 1861, *McClellan Papers,* 100; Theodore C. Pease and James G. Randall, eds., *The Diary of Orville Hickman Browning,* 2 vols. (Springfield: Illinois State Historical Library, 1925, 1933), 1:563; Fishel, "Pinkerton and McClellan," 135.

13. McClellan to Mrs. McClellan, Aug. 2, 8, 1861; to Lincoln, Aug. 2, 1861, *McClellan Papers,* 75, 81, 71–75; Dennett, ed., *Hay Diaries,* 33.

14. Bancroft to Mrs. Bancroft, Dec. 16, 1861, M. A. DeWolfe Howe, ed., *The Life and Letters of George Bancroft,* 2 vols. (New York: Charles Scribner's Sons, 1908), 2:47; Samuel S. Blair to James H. Bell, Dec. 18, 1861, Bell Papers, Duke University.

15. Lincoln to McClellan, c. Dec. 1, 1861, Roy P. Basler et al., ed., *The Collected Works of Abraham Lincoln,* 9 vols. (New Brunswick: Rutgers University Press, 1953–55), 5:34; McClellan to Lincoln, Dec. 10, 1861, *McClellan Papers,* 143.

16. Montgomery Meigs, "General M. C. Meigs on the Conduct of the Civil War," *American Historical Review,* 26:2 (Jan. 1921): 292–93; "Memorandum of General McDowell," Henry J. Raymond, *The Life and Public Services of Abraham Lincoln* (New York: Derby and Miller, 1865), 773; Malcolm Ives to James Gordon Bennett, Jan. 15, 1862, Bennett Papers, Library of Congress; William B. Franklin in A. K. McClure, ed., *The Annals of the War* (Philadelphia: Times Publishing, 1879), 79.

17. Lincoln to Buell and Halleck, Jan. 13, 1862, Basler, ed., *Collected Works of Lincoln,* 5:98–99.

18. Stanton to Charles A. Dana, Jan. 24, 1862, Dana, *Recollections of the Civil War: With the Leaders at Washington and in the Field in the Sixties* (New York: D. Appleton, 1898), 5.

19. Lincoln memorandum; President's General War Order no. 1; President's Special War Order No. 1, Basler, ed., *Collected Works of Lincoln,* 5:119, 111–12, 115; John Codman Ropes, *The Story of the Civil War,* 2 vols. (New York: G. P. Putnam's Sons, 1895–98), 1:226.

20. George B. McClellan, *Report on the Organization of the Army of the Potomac, and of Its Campaigns in Virginia and Maryland* (Washington: GPO, 1864), 43; McClellan to Stanton, Feb. 3, 1862, *McClellan Papers,* 162–70; Lincoln to McClellan, Feb. 3, April 9, 1862, Basler, ed., *Collected Works of Lincoln,* 5:118–19, 185.

21. President's General War Order No. 3; President's War Order No. 3; Lincoln to McClellan, March 13, 1862, Basler, ed., *Collected Works of Lincoln,* 5:151, 155, 157–58; McClellan to Barlow, March 16, 1862; to Lorenzo Thomas, April 1, 1862; to Mrs. McClellan, April 1, 1862, *McClellan Papers,* 213, 222–23, 223.

22. Council of war report, March 13, 1862, *OR,* ser. 1, vol. 11, pt. 3, p. 58; Stanton to Lincoln, March 30, 1862, Stanton Papers, Library of Congress; Charles Sumner to John A. Andrew, May 28, 1862, Andrew Papers, Massachusetts Historical Society.

23. McClellan to Stanton, March 18, 1862, *McClellan Papers,* 214; Heintzelman diary, July 26, 1862, Heintzelman Papers, Library of Congress.

24. Lincoln to McClellan, April 9, 1862, Basler, ed., *Collected Works of Lincoln,* 5:184–85; McClellan to Mrs. McClellan, April 8, 1862, *McClellan Papers,* 234.

25. Dabney H. Maury, *Recollections of a Virginian in the Mexican, Indian, and Civil Wars* (New York: Charles Scribner's Sons, 1894), 161.

26. McClellan to Lincoln, July 7, 1862; to Mrs. McClellan, July 9, 1862, *McClellan Papers,* 344–45, 348.

27. Lincoln to McClellan, July 13, 1862, Basler, ed., *Collected Works of Lincoln,* 5:322; McClellan to Lincoln, Aug. 29, 1862, *McClellan Papers,* 416; Adams S. Hill to Sydney H. Gay, Aug. 31, 1862, Gay Papers, Columbia University; Dennett, ed., *Hay Diaries,* 45.

28. Bates memorandum, Sept. 2, 1862, Abraham Lincoln Papers, Library of Congress; Howard K. Beale, ed., *Diary of Gideon Welles,* 3 vols. (New York: Norton 1960), 1:105, 113, 116; Dennett, ed., *Hay Diaries,* 47.

29. McClellan to Lincoln, May 30, 1861, July 7, 1862; to Mrs. McClellan, c. Nov. 14, 1861, *McClellan Papers,* 28, 344, 132; Key to Stanton, June 16, 1862, *OR,* ser. 1, vol. 11, pt. 1, pp. 1052–56; McClellan to William H. Aspinwall, Sept. 26, 1862; to Mrs. McClellan, Sept. 25, 1862, *McClellan Papers,* 482, 481; *The Round Table,* March 12, 1864; *Philadelphia Press,* Oct. 13, 1863.

30. McClellan to Mrs. McClellan, Sept. 22, Oct. 5, 1862, *McClellan Papers,* 477, 490; John G. Nicolay and John Hay, *Abraham Lincoln: A History,* 10 vols. (New York: Century, 1886–90), 6:175.

31. Lincoln to McClellan, Oct. 13, 1862, Basler, ed., *Collected Works of Lincoln,* 5:460–61; McClellan to Lincoln, Oct. 17, 1862, *McClellan Papers,* 499; Darius Couch in *Battles and Leaders of the Civil War,* eds., Robert U. Johnson and Clarence C. Buel, 4 vols. (New York: Century, 1887–88), 3:105–6.

32. McClure to Mrs. McClellan, Jan. 13, 1892, McClellan Papers.

Two: Wilderness and the Cult of Manliness: Hooker, Lincoln, and Defeat

I want to thank Matthew Noah Vosmeier and Sarah McNair Vosmeier, graduate students in American history at Indiana University, for helpful comments on an earlier version of this article. Sylvia Neely also read and commented usefully—and read aloud from guidebooks as in May 1993 we trudged over the Virginia battlefields mentioned in this article.

1. Walter H. Hebert, *Fighting Joe Hooker* (Indianapolis: Bobbs-Merrill, 1944), 41.

2. The Abraham Lincoln Papers at the Library of Congress for the month following Hooker's appointment do not contain a single letter concerning it. As for the general's politics, Lincoln noted in his famous letter that Hooker kept them separate from his military career. Hooker evidently supported Lincoln's re-election in 1864 but after the war said, "I was never anything else than a Democrat." The best discussion of his appointment appears in Kenneth P. Williams, *Lincoln Finds a General: A Military Study of the Civil War,* 2 vols. (New York: Macmillan, 1949), 2:547-551, 829n. Political interpretations of military appointments in the Civil War thrived in the days of T. Harry Williams's greatest influence. For an excellent statement of the idea that politics influenced the highest military appointments, see Stephen E. Ambrose, *Halleck: Lincoln's Chief of Staff* (Baton Rouge: Louisiana State University Press, 1962), 102; see also Hebert, *Fighting Joe Hooker,* 164, 290, 295, and Williams himself in *Lincoln and the Radicals* (Madison: University of Wisconsin Press, 1941), 270–73.

3. Henry W. Raymond, ed., "Extracts from the Journal of Henry J. Raymond. II," *Scribner's Monthly,* 19 (Jan. 1880), 422; Raymond, ed., "Extracts . . . III," *Scribner's Monthly,* 19 (Feb. 1880), 705.

4. Ambrose, *Halleck,* 81, 93.

5. Brison D. Gooch, "The Crimean War in Selected Documents and Secondary Works Since 1940," *Victorian Studies,* 1 (March 1958): 271. American Victorians may have shared this cultural trait of lack of ability at war with their English cousins. The classic account of the corruption of the British officer corps and its effects on the Crimean War is Cecil Woodham-Smith, *The Reason Why* (orig. pub. 1953; New York: Dutton, 1960).

6. On the French see Brison D. Gooch, *The New Bonapartist Generals in the Crimean War: Distrust and Decision-Making in the Anglo-French Alliance* (The Hague: M. Nijhoff, 1959).

7. On Prussia see Michael Howard, *The Franco-Prussian War: The German Invasion of France, 1870–1871* (orig. pub. 1961; New York: Methuen, 1985), 20–29, 60–63.

8. Louis Gottschalk, *Lafayette and the Close of the American Revolution* (2nd ed.; Chicago: University of Chicago Press, 1965), 233; see also 235.

9. Among other virtues. Faith in democracy and a fourth virtue, which might be called recognition of the nature of "total war" and application of its insights to the Civil War, cannot be addressed here.

10. Roy P. Basler et al., eds., *The Collected Works of Abraham Lincoln,* 9 vols. (New Brunswick, N.J.: Rutgers University Press, 1953–55), 6:79.

11. Quoted in G. S. Boritt, *Lincoln and the Economics of the American Dream* (Memphis, Tenn.: Memphis State University Press, 1978), 271.

12. Horace Porter, *Campaigning with Grant* (orig. pub. 1897; New York: Bantam Books, 1991), 123.

13. Tyler Dennett, ed., *Lincoln and the Civil War in the Diary and Letters of John Hay* (orig. pub. 1939; New York: Da Capo, 1988), 176.

14. On Virginia terrain see Richard M. McMurry, *Two Great*

Rebel Armies: An Essay in Confederate Military History (Chapel Hill: University of North Carolina Press, 1989).

15. *The War of the Rebellion: A Compilation of the Official Records of the Union and Confederate Armies,* 127 vols. (Washington: GPO, 1880–1901), ser. 1, vol. 25, pt. 1, p. 193. Hereafter *OR*.

16. On dysfunctional military technology see Paddy Griffith, *Battle Tactics in the Civil War* (orig. pub. 1987; New Haven: Yale University Press, 1989). On the surprisingly insignificant role of technology in the land battles of the Civil War see Robert V. Bruce, *Lincoln and the Tools of War* (orig. pub. 1956; Urbana: University of Ilinois Press, 1989) and *The Launching of Modern American Science, 1846–1876* (orig. pub. 1987; Ithaca: Cornell University Press, 1988).

17. *OR,* ser. 1, vol. 25, pt. 1, pp. 195–97.

18. Basler, ed., *Collected Works of Lincoln,* 5:355; 2:111; 5:185; Ulysses S. Grant, *Memoirs and Selected Letters* (New York: Library of America, 1990), 473.

19. Basler, ed., *Collected Works of Lincoln,* 6:64, 165.

20. Ibid., 5:460–461; *OR,* ser. 1, vol. 25, pt. 2, p. 246.

21. Ernest B. Furgurson, *Chancellorsville 1863: The Souls of the Brave* (New York: Alfred A. Knopf, 1992), 140–41.

22. When Michael C. C. Adams came to write a new introduction to his wonderful book, originally entitled *Our Masters the Rebels,* in 1992, fourteen years after its original publication, he admitted, "Were I writing this work now, I should pay more attention to" gender studies. See Adams, *Fighting for Defeat: Union Military Failure in the East, 1861–1865* (Lincoln: University of Nebraska Press, 1992). He thus expressed the regret and afterthought of a generation of new Civil War historians. Such studies remain in their infancy, and definitions of terms are hardly firm. "Manly" often meant "uncomplaining" in the middle of the nineteenth century. Most modern writers emphasize the term's connotation of self-restraint and, to a lesser degree, independence. See

for example, Reid Mitchell, *The Vacant Chair: The Northern Soldier Leaves Home* (New York: Oxford University Press, 1993), 3–18, 170–72. See also David W. Blight, "No Desperate Hero: Manhood and Freedom in a Union Soldier's Experience," in Catherine Clinton and Nina Silber, eds., *Divided Houses: Gender and the Civil War* (New York: Oxford University Press, 1992) 63–67, and in the same volume, Jim Cullen, " 'I's a Man Now': Gender and African American Men," 79. In the *Oxford English Dictionary* courage is the first term associated with "manly," and my article focuses on the traditional association with that virtue women and children were thought not to have. Moreover, it focuses on that idea in the thrusting, positive, and assertive sense explored by Gerald F. Linderman, *Embattled Courage: The Experience of Combat in the American Civil War* (New York: Free Press, 1987).

23. G. S. Boritt, "War Opponent and War President," in Boritt, ed., *Lincoln the War President: The Gettysburg Lectures* (New York: Oxford University Press, 1992), 191.

24. See, for example, C. Vann Woodward and Elisabeth Muhlenfeld, eds., *The Private Mary Chestnut: The Unpublished Civil War Diaries* (New York: Oxford University Press, 1984), 22, 34.

25. Mark E. Neely, Jr., *The Last Best Hope of Earth: Abraham Lincoln and the Promise of America* (Cambridge: Harvard University Press, 1993), 186.

26. John Keegan, *The Mask of Command* (New York: Viking, 1987), 168–234.

27. *OR,* ser. 1, vol. 25, pt. 2, p. 119.

28. Stephen W. Sears, *George B. McClellan: The Young Napoleon* (New York: Ticknor & Fields, 1988), 196; *OR,* ser. 4, vol. 3, pp. 104–5.

29. James M. McPherson, *Battle Cry of Freedom: The Civil War Era* (New York: Oxford University Press, 1988), 330.

30. Jonathan Letterman in *Medical Recollections of the Army of the Potomac* (New York: D. Appleton, 1866), 125–26, noted: "On the afternoon of May 2d this building came within range of the enemy's guns, planted on his left, centre, and right, being the centre of a converging fire—a location for which Commanding Generals of the Army of the Potomac seemed to have a peculiar partiality. Previous to this time I had all the wounded, who could be removed, taken further to the rear, and directed that no more should be received in the house, as it was evident, from the position of our line, that when the battle opened in earnest, the building would also come within range of his musketry."

31. *OR,* ser. 1, vol. 19, pt. 1, p. 218.

32. Hebert, *Fighting Joe Hooker,* 204.

33. Samuel P. Bates, *The Battle of Chancellorsville* (orig. pub. 1882: Gaithersburg, Md.: Ron R. Van Sickle Military Books, 1987), 42.

34. *OR,* ser. 1, vol. 25, pt. 2, p. 241.

35. Furgurson, *Chancellorsville 1863,* 335n. This was, of course, Hooker's view of Howard, not the real Howard.

36. *OR,* ser. 1, vol. 25, pt. 2, pp. 377–78.

37. Francis A. Walker, *History of the Second Army Corps in the Army of the Potomac* (New York: Charles Scribner's Sons, 1886), 253–54.

38. Furgurson, *Chancellorsville 1863,* 181.

39. Andrew A. Humphreys, *The Virginia Campaign of 1864 and 1865* (orig. pub. 1885; New York: Charles Scribner's Sons, 1908), 55.

40. *OR,* ser. 1, vol. 25, pt. 1, p. 196. The Confederate colonel David Gregg McIntosh observed of Hooker, "He was now on the ground which Grant had to fight for in the succeeding year, with the differencc that then Grant was east of Lee, and now Hooker was west of Lee." See McIntosh, *The Campaign of Chancellorsville* (Richmond: William Ellis Jones' Sons, Printers, 1915),

20. In one of the earliest books on the battle, Jedediah Hotchkiss and William Allan's *The Battle-Fields of Virginia: Chancellorsville* . . . (New York: D. Van Nostrand, 1867), the maps, a principal feature of the work, carried the title " 'The Wilderness' or Chancellorsville, Salem Church, and Fredericksburg."

41. Augustus Choate Hamlin, *The Battle of Chancellorsville: The Attack of Stonewall Jackson and His Army upon the Right Flank of the Army of the Potomac . . . on Saturday Afternoon, May 2, 1863* (Bangor, Maine: privately pub., 1896), 13–14, 8.

42. Olivier Fraysée, *Lincoln, Land, and Labor* (Urbana: University of Illinois Press, 1994).

Three: "Unfinished Work": Lincoln, Meade and Gettysburg

Friends kind enough to read this chapter include Robert V. Bruce, Gary Gallagher, William Ridinger, Col. Jake Sheads and John Y. Simon. I am specially indebted to Kent Masterson Brown, the expert on Lee's retreat from Gettysburg, for several long and generous conversations. He convinced me about the length of the Confederate line at Williamsport. The germ of the idea contained in the title of this chapter came from Jake Fisher of Trinity College, who transferred to Gettysburg for the spring of 1993 to take my Lincoln Seminar. I also benefited from discussing the Meade-Lincoln relationship with the other members of the seminar, nearly all of whom looked with much favor on the general: Erik Breshnehan, Rich Butler, Erin Cahil, Steve Grow, Mark Kolhoff, Chris Lyerly, Jim Newman, and Al Pennino. Four other students, Melissa Becker, Susan Fiedler, Deborah Huso, and Christopher Patterson, ably served as my assistants. Professor Michael Burlingame generously shared materials on Lincoln's lost order, convinced me to look at it seriously, and called to my attention the fact that the printed edition of John Hay's diary is inaccurate.

Finally, I tramped the routes and fields of Lee's retreat at various times in the good company of John Schildt, Edwin Besch, Ted Alexander, and Pete Vermilyea. Meade's pursuit I followed mostly alone.

1. Roy P. Basler, ed., Marion Dolores Pratt and Lloyd A. Dunlap, assist. eds., *The Collected Works of Abraham Lincoln,* 9 vols. (New Brunswick: Rutgers University Press, 1953–55), 6:323.

2. Ibid., 6:314. Lincoln signed but probably did not compose the text.

3. Ibid., 327.

4. Memorandum of Rush C. Hawkins, Aug. 17, 1872, Hawkins Papers, John Hay Library, Brown University; George H. Thacher, "Lincoln and Meade After Gettysburg," *American Historical Review,* 32 (1927): 282–83; Diary of John Hay, July 15, 1863, Hay Papers, Brown University.

5. Captain George Meade to Mrs. George Gordon (Margaretta Sergeant) Meade, July 7, 1863, Meade Papers, Historical Society of Pennsylvania; George Gordon Meade to H. W. Halleck, July 9, 1863, *The War of the Rebellion: A Compilation of the Official Records of the Union and Confederate Armies,* 127 vols., index, and atlas (Washington: GPO, 1880–1901), ser. 1, vol. 27, pt. 1, pp. 86–87; Meade to Mrs. Meade, July 16, 1863, Meade Papers.

6. Basler, ed., *Collected Works of Lincoln,* 6:283, 311; 5:460–61.

7. Ibid., 5:501; 6:341; Benjamin Brown French, *Witness to the Young Republic: A Yankee Journal, 1828–1870,* Donald B. Cole and John J. McDonough, eds., (Hanover, N.H.: University Press of New England, 1989), 424; Howard K. Beale, ed., Alan W. Brownsword, assist., *Diary of Gideon Welles, Secretary of Navy Under Lincoln and Johnson,* 3 vols. (New York: Norton, 1960), 1:344.

8. Ida Tarbell's Interview with A. B. Chandler, Sept. 16, 1898, Ida Tarbell Papers, Allegheny College; George C. Gorham, *Life and Public Services of Edwin M. Stanton,* 2 vols. (Boston: Houghton, 1899), 2:99.

9. Meade to Mrs. Meade, June 29, 1863, Meade Papers.

10. Ibid., Aug. 16, 1862, Jan. 2 and May 8, 1863, Meade Papers.

11. Ibid., Dec. 2, 1846, Meade Papers. Meade spoke about himself in the context of the regular army's opinions about volunteer soldiers. Someone corrected the spelling of "miliau" to "milieu."

12. Basler, ed., *Collected Works of Lincoln,* 6:314, 320.

13. Meade to Mrs. Meade, July 5, 1863, Meade Papers.

14. *OR,* ser. 1, vol. 27, pt. 3, p. 519.

15. David H. Bates, *Lincoln in the Telegraph Office* (New York: Century, 1907), 155–56; Allen Thorndike Rice, ed., *Reminiscences of Abraham Lincoln by Distinguished Men of His Time* (New York: North American Review, 1888), 402; Hay Diary, July 14, 1863, Hay Papers.

16. Gary W. Gallagher, ed., *Fighting for the Confederacy: the Personal Recollections of General Edward Porter Alexander* (Chapel Hill: University of North Carolina Press, 1989), 267.

17. James H. Campbell to Juliet Campbell, June [sic., July] 5, 1863, Clements Library, University of Michigan, Ann Arbor.

18. John D. Imboden, "The Confederate Retreat from Gettysburg," *Battles and Leaders of the Civil War,* Robert Underwood Johnson and Clarence Clough Buell, eds., 4 vols. (New York: Century, 1988), 3:427.

19. Basler, ed., *Collected Works of Lincoln,* 6:318.

20. *Welles Diary,* 1:363.

21. Basler, ed., *Collected Works of Lincoln,* 6:319; James J. Smart, ed., *A Radical View: The "Agate" Dispatches of Whitelaw*

Reid, 1861–1865, 2 vols. (Memphis: Memphis State University Press, 1976), 2:62.

22. Basler, ed., *Collected Works of Lincoln,* 6:319–20.

23. Gallagher, ed., *Recollections of Alexander,* 270–71.

24. Ibid., 270.

25. Imboden, *Battles and Leaders,* 3:429; Robert E. Lee to Mrs. Mary Custis Lee, July 12, 1863, Clifford Dowdey, ed., Louis H. Manarin, assoc. ed., *The Wartime Papers of R. E. Lee* (Boston: Little, Brown, 1961), 547; *OR,* ser. 1, vol. 27, pt. 2, p. 302.

26. Meade to Mrs. Meade, July 8, 1863, Meade Papers.

27. Basler, ed., *Collected Works of Lincoln,* 6:322.

28. *OR,* ser. 1, vol. 27, pt. 1, p. 86; David W. Blight, ed., *When This War is Over: The Civil War Letters of Charles Harvey Brewster* (Amherst: University of Massachusetts Press, 1992), 241; William Garrett Piston, ed., " 'The Rebs Are Yet Thick About Us': The Civil War Diary of Amos Stouffer of Chambersburg," *Civil War History,* 38 (1992), 223.

29. *Welles Diary,* 1:363; Tarbell interview with Chandler; Basler, ed., *Collected Works of Lincoln,* 6:321.

30. Hawkins Memorandum.

31. Basler, ed., *Collected Works of Lincoln,* 7:282.

32. Thirteen years after General Hawkins recorded the presidential order, Robert wrote in his own hand: "My father then said that he at once sent an order to General Meade . . . directing him to attack Lee's army with all his force immediately, and that if he was successful in the attack he might destroy the order but if he was unsuccessful he might preserve it for his vindication." Memorandum, January 5, 1885, enclosed with Robert Lincoln to John G. Nicolay, January 5, 1885, John G. Nicolay Papers, Library of Congress. On April 17, 1897, former Iowa Senator James Harlan, Robert's father-in-law, gave this version: "The President sent an order, privately, directing General Meade to follow up his victory by an immediate attack on Lee's retreating army, and thus, if possible,

prevent the re-crossing of the Potomac by the Confederate forces accompanied by a confidential letter authorizing him to make the order public in case of disaster and in case of success to destroy both the order and confidential letter. This much you may rely upon as historically true." Isaac Newton Phillips, *Abraham Lincoln* (Bloomington, Ill., 1901), 46 n. The source of Senator Harlan's information is obvious. Early in the twentieth century a golfing partner of Robert wrote down President Lincoln's words "immediately" after they had been related by his son. "I at once wrote Meade to attack without delay, and if successful to destroy my letter, but in case of failure to preserve it for his vindication." Thacher, "Lincoln and Meade after Gettysburg," 282. It is not inconceivable that Robert Lincoln confused his father's lost order with another, like message incorporated in a longer letter to Halleck, October 16, 1863, and then sent by him to Meade. Basler, ed., *Collected Works of Lincoln,* 6:518. However, the traumatic moment of Robert seeing his father cry and being told of its cause etched itself deeply into the son's memory. His discreet and essentially unchanging repetition of this account, from a young age to his death, makes it credible.

A less than well-informed F. Lauriston Bullard discussed the matter of Lincoln's lost order in print: "President Lincoln and General Meade After Gettysburg," Parts 1 and 2, *Lincoln Herald,* 47 (1945), 1:30–34; 3–4:13–16. To follow further the trail of Lincoln's lost order, see in addition to the other items mentioned in the footnotes: Robert T. Lincoln to John G. Nicolay, June 14, 1878; Nicolay to Lincoln, June 25, 1878, and Sept. 5, 1881, Nicolay Papers, Library of Congress; Robert T. Lincoln to Abner Doubleday, Sept. 29, 1881; and to Isaac N. Arnold, March 27, 1884, Robert Todd Lincoln Papers, Letter Press Book 5 and 10, Illinois State Historical Society; Robert T. Lincoln to Isaac N. Arnold, March 27, 1884, Abraham Lincoln Collection, Chicago Historical Society; Newton McMillan, "Memoirs of Lincoln," Portland *Oregonian,* March 28, 1895 (cf. Ida Tarbell Interview of Joseph Medill, June 28, 1895), Tarbell Papers, Allegheny College; Isaac N. Phillips to Charles W. McClellan, Oct. 19, 1908, enclosing annotated excerpt from Phillips, "Abraham Lincoln," McClellan Collection, John Hay Library, Brown University; R. Lincoln to Helen Nicolay, May 29, 1912, enclosing annotated text, Nicolay Papers, Library of Congress; R. Lincoln to Isaac Markens, Apr. 6, 1918, Robert Todd Lincoln Collection, Chicago Historical Society; Robert T. Lincoln in April 27, 1925, enclosing H. L. to [Roger L.] Scaife, n.d., Albert J. Beveridge Papers, Library of Congress. More research is needed in primary sources.

33. Robert T. Lincoln to Isaac Arnold, Nov. 11, 1883, Arnold Papers, Chicago Historical Society. A copy of the letter is in the Robert Todd Lincoln Papers, Letter Press Book 9, Illinois State Historical Society.

34. George Meade to Mother, July 11, 1863, Meade Papers. Newspaperman Noah Brooks later recalled meeting the Vice President, after Lee's escape: "Hamlin raised his hands and turned away his face with a gesture of despair." Noah Brooks, *Washington, D.C. in Lincoln's Time,* Herbert Mitgang, ed. (Chicago: Quadrangle, 1971), 92; cf. McMillan, "Memories of Lincoln"; and Hannibal Hamlin to Emery Hamlin, July 6, 1863, Hamlin Papers, University of Maine.

35. Thacher, "Lincoln and Meade after Gettysburg," 282.

36. Hay Diary, July 14, 11, 13, 1863.

37. *OR,* ser. 1, vol. 27, pt. 1, p. 91.

38. Ibid., 92.

39. Justus Scheibert, *Seven Months in the Rebel States During the North American War, 1863* (Confederate Publishers: Tuscaloosa, Alabama, 1958), 122.

40. Hay Diary, July 14, 1863. Chandler remembered Lincoln's saying: "Yes. He will fight a magnificent battle when there is no enemy there." Tarbell interview with Chandler.

41. James Longstreet, *From Manassas to Appomattox* (Philadelphia, 1896), 429.

42. Moxley G. Sorrel, *Recollections of a Confederate Staff Officer* (New York, 1905), 171–2; Coddington, *The Gettysburg Campaign,* 572.

43. *Welles Diary,* 1:370.

44. Ibid.; Hay Diary, July 14, 1863.

45. *Welles Diary,* 1:369, 371–72.

46. Meade to Mrs. Meade, May 8, Oct. 12, 1863, Meade Papers; T. Harry Williams, *Lincoln and His Generals* (New York: Knopf, 1952), 267.

47. William O. Stoddard, *Inside the White House in Wartimes* (New York: C. L. Webster, 1890), 178–79; George Meade to Laycee [?], July 17, 1863, Meade Papers.

48. George B. McClellan to Meade, July 11, 1863, Meade Papers; Hunt to Alexander Webb, Jan. 19, 1888, Alexander Webb Papers, Yale University.

49. James Cornell Biddle to Mrs. Gertrude Biddle, July 8, 11, 14, 16, 18, 1863, James C. Biddle Papers, Historical Society of Pennsylvania. Newspaperman Noah Brooks, with the army at the time, recalled that "some of the younger men at headquarters did not hesitate to say that Meade had been "most egregiously fooled." Brooks, *Washington in Lincoln's Time*, 89.

50. Ibid., 83; *OR,* ser. 1, vol. 27, pt. 1, p. 92; Meade to Mrs. Meade, July 16, 18, 1863, Meade Papers.

51. Basler, ed., *Collected Works of Lincoln,* 6:327–9.

52. Meade to Mrs. Meade, July 21, 1863, Meade Papers.

53. *OR,* ser. 1, vol. 27, pt. 1, p. 109; Meade to Mrs. Meade, July 21, 1863, Meade Papers.

54. Basler, ed., *Collected Works of Lincoln,* 6:341.

55. Hay Diary, July 19, 1863; Roy P. Basler, ed., *The Collected Works of Abraham Lincoln, Supplement, 1832–1865* (Westport, Ct.: Greenwood, 1974), 194.

56. Basler, ed., *Collected Works of Lincoln,* 6:509–10, 518; *OR,* ser. 1, vol. 29, pt. 2, pp. 293, 333.

57. *St. Paul and Minneapolis Pioneer Press,* Dec. 7, 1884, cited in Williams, *Lincoln and his Generals,* 271; Meade to Mrs. Meade, Oct. 23, Nov. 3, 1863, Meade Papers.

58. *Welles Diary,* 1:439.

59. Meade to Mrs. Meade, Dec. 2, 1863.

60. Basler, ed., *Collected Works of Lincoln,* 6:354 (cf. 5:460–61); Biddle to wife, July 18, 1863, Biddle Papers.

61. George Meade, *Life and Letters of George Gordon Meade,* 2 vols. (New York: Charles Scribner's Sons, 1913), 2:174; Basler, ed., *Collected Works of Lincoln,* 7:273; Meade to Mrs. Meade, Aug. 27, 31, 1863, Meade Papers.

62. Edward Everett Diary, Nov. 19, 1863, Massachusetts Historical Society.

63. Basler, ed., *Collected Works of Lincoln,* 7:20.

64. June 28, 1865, Meade Papers.

65. Samuel Ensminger to his mother (Mary Ann Dewalt Ensminger) and brother (Joseph Ensminger), May 3, 1863, and to wife and family (Magdalena Myers Ensminger and Amanda Ensminger), February 14, 1863, in the possession of Clarence M. Swinn, Jr., Gettysburg. I am also indebted to Mr. Swinn for his generous sharing of information about the Ensminger family. Mr. Swinn's father was the foster child of Sergeant Ensminger's daughter, Amanda.

66. Ensminger to wife, July 14, 1863, Swinn Papers.

67. John D. Billings, *Hardtack and Coffee* (Boston: Smith, 1887), 71–72. Lines are repeated as they would have been sung. Billings wrote that some of the stanzas "may have got distorted with the lapse of time. But they are essentially correct." A song is a fine, if complicated, shortcut to the heart of the Army of the Potomac. But a song is no substitute for the systematic study of the views of the common soldiers. That task remains for the future.

Four: Lincoln and Sherman

1. Richard Berringer, Herman Hattaway, Archer Jones, and William N. Still, Jr., *Why the South Lost the Civil War* (Athens: University of Georgia Press, 1986); Mark E. Neely, Jr., *The Fate of Liberty: Abraham Lincoln and Civil Liberties* (New York: Oxford University Press, 1991). Also see my discussion of Neely in "The Illiberal Lincoln," *Canadian Review of American Studies,* 23 (Winter, 1993): 195–201.

2. John Sherman to W. T. Sherman, Washington, Jan. 2, 1862, May 7, July 18, 1863, July 25, Sept. 4, 1864, Sherman Papers, Library of Congress (hereafter LC).

3. For fuller readings of Sherman's politics and personality, here and elsewhere in this essay, one will have to await my study, *Citizen Sherman* (New York: Random House, forthcoming).

4. Sherman, *Memoirs of General William T. Sherman,* reprinting of 1st ed. (New York: Da Capo, 1984), 1:168. Lincoln's denial of the reality of the onrushing war which he expressed to Sherman was psychological as well political. See Gabor S. Boritt, "'And the War Came?' Abraham Lincoln and the Question of Individual Responsibility," in Gabor Boritt, ed., *Why the Civil War Came* (New York: Oxford University Press, forthcoming).

5. Sherman to John Sherman, Lancaster, March 9, Cincinnati, March 21, St. Louis, April 18, 22, 26, 1861, Sherman Papers, LC.

6. Sherman to Montgomery Blair, St. Louis, April 18, 1861, reprinted in *Memoirs,* 1:170–71; Sherman to John Sherman, Cincinnati, March 21, St. Louis, April 22, 1861, Sherman Papers, LC.

7. Sherman, Report, Ft. Corcoran, July 25, 1861, *The War of the Rebellion: Official Records of the Union and Confederate Armies,* 127 vols. (Washington: GPO, 1880–1901), ser. 1, vol. 2, pp. 367–71 (hereafter *OR*); Sherman to Adjutant-General, Fort

Corcoran, July 22, 1861, *OR,* vol. 2, p. 755; Sherman to Ellen Sherman, Washington, Aug. 3, 1861, in Mark A. DeWolfe Howe, ed., *Home Letters of General Sherman* (New York: Charles Scribner's Sons, 1909), 212–13.

8. Ibid.; Sherman to John Sherman, Ft. Gore, Aug. 19, 1861, Sherman Papers, LC, reprinted in part in Rachel Sherman Thorndike, ed., *The Sherman Letters: Correspondence Between General and Senator Sherman from 1837 to 1891* (1894, reprinted, New York: Da Capo, 1969), 126–27.

9. Sherman, *Memoirs,* 1:188–91.

10. Ibid., 1:192–93; Sherman to Ellen Sherman [c. Aug. 20, 1861], Sherman Family Papers, Notre Dame University Archives (hereafter ND).

11. Ellen Sherman to Abraham Lincoln, Lancaster, Dec. 19, 1861, copies in ND, and Sherman Papers, LC.

12. Thomas Ewing to John Sherman, Lancaster, Dec. 22; Hugh Ewing to Sherman, Washington, Dec. 23, 1861; Ellen Sherman to Sherman, Lancaster, Jan. 11, Feb. 4, Washington, Jan. 29, 1862, all in ND. Emphasis in the original.

13. Ellen Sherman to Sherman, Washington, Jan. 29, 1862, ND; Anna McAllister, *Ellen Ewing: Wife of General Sherman* (New York: Benziger Brothers, 1936), 208–10.

14. Lincoln to James C. Conkling, Washington, Aug. 26, 1863, in Roy Basler et al., eds., *The Collected Works of Abraham Lincoln,* 9 vols. (New Brunswick: Rutgers University Press, 1953), 6:409–10.

15. Sherman to Ellen Sherman, Camp Opposite Vicksburg, April 17, 1863, *Sherman Letters,* 251–52; Sherman to Halleck, Memphis, Oct. 10, 1863, *OR,* vol. 30, pt. 4, pp. 234–35.

16. Sherman to John, Camp Before Vicksburg, April 26, 1863, Sherman Papers, LC.

17. Lorenzo Thomas to Sherman, Natchez, March 30, 1864; [Sherman]. Special Field Orders #16, In the Field Near Dallas,

Georgia, June 3, 1864, *OR,* ser. 3, vol. 4, pp. 210–11, 432. It is worth nothing that even while Sherman was impeding black recruitment, his step-brother/brother-in-law, Thomas Ewing, Jr., was organizing black regiments in Missouri. Ibid., 433.

18. Lorenzo Thomas to Sherman, Nashville, June 19, 1864, *OR,* ser. 3, vol. 4, p. 437; Sherman to Thomas, In the Field, Big Shanty, June 21, 1864, *OR,* vol. 39, pt. 2, pp. 132–33; Sherman to Thomas, Near Kenesaw Mountain, June 26, 1864, *OR* ser. 3, vol. 4, pp. 454–55; Sherman to Halleck, near Chattahoochee Creek, July 14, 1864, *OR,* vol. 38, pt. 5, p. 137.

19. Lincoln to Sherman, Washington, July 18, 1864, Basler, ed., *Collected Works of Lincoln,* 6:449–50.

20. Lincoln to Isaac M. Schermerhorn, Washington, Sept. 12, 1864, Basler, ed., *Collected Works of Lincoln,* 8:2.

21. Sherman to Lincoln, near Atlanta, July 21, 1864, *OR,* vol. 38, pt. 5, p. 210.

22. Sherman to John A. Spooner, In the Field Near Atlanta, July 30, 1864, *OR,* vol. 38, pt. 5, pp. 305–6.

23. Sherman to William M. McPherson [Atlanta, c. Sept. 1864], Sherman Papers, Huntington Library, San Marino, California.

24. Sherman to Halleck, South to Atlanta, Sept. 4, 1864, *OR,* vol. 38, pt. 5, pp. 791–95.

25. Levi D. Bryant to his wife, March 28, 1865, Michael Winey Collection, United States Army Military History Institute, Carlisle Barracks, Pennsylvania, quoted in Joseph T. Glatthaar, *The March to the Sea and Beyond: Sherman's Troops in the Savannah and Carolinas Campaigns* (New York: New York University Press, 1985), 67. Glatthaar has an excellent discussion of the aversive reactions of Sherman's men to the introduction of black troops, at pages 66–67.

26. Sherman, *Memoirs,* 2:250–53.

27. Reprinted in Basler, ed., *Collected Works of Lincoln,*

8:332–33. Lincoln's texts were St. Matthew 18:7, and Psalms 19:9.

28. Sherman, *Memoirs,* 2:249.

29. Sherman to Halleck, Summerville, Ga., Oct. 18; Sherman to James H. Wilson, Oct. 19; Sherman to George Thomas, Oct. 20, Rome, Oct. 29; Sherman to Grant, Kingston, Nov. 6, *OR,* vol. 39, pt. 3, pp. 358, 377–78, 498, 659–61.

30. Sherman to Lincoln, Savannah, Dec. 22; Lincoln to Sherman, Washington, Dec. 26, 1864, *OR,* vol. 44, pp. 783, 809.

31. Sherman and Johnston, Memorandum of Agreement, Near Durham's Station, North Carolina, April 18, 1865, *OR,* vol. 47, pt. 3, pp. 243–45.

32. Sherman, *Memoirs,* 2:327; Sherman to Hon. J. N. Arnold, Washington, Nov. 28, 1872, Sherman Papers, LC; Sherman to Ellen Sherman, Goldsborough, April 5, 1865, *Sherman Letters,* 338–42. This is an absolutely central point, as both Sherman and his 1931 biographer Lloyd Lewis used the dead Lincoln to justify their own reactionary Reconstruction politics. Lincoln wanted to forgive the South and act against the Radicals, who were eager for punishment, the story goes, and Reconstruction proved terribly unfair to Southern whites, and demonstrated far too much concern with the undeserving blacks. This was a projection of values onto Lincoln, who unlike his successor, Andrew Johnson, was a master of keeping political options open, and who had grown in the direction of the Radicals on racial and political issues throughout the war. To prove his thesis, in his *Memoirs,* Sherman even had Admiral David Porter, who was there, back up his political interpretation of the March 27, 28, 1865, meeting. Sherman, *Memoirs,* 2:324–32; Lewis, *Sherman: Fighting Prophet* (New York: Harcourt, Brace, 1931), 518–80.

Five: Grant, Lincoln, and Unconditional Surrender

1. Ulysses S. Grant to Julia Dent Grant, July 1, 1852, John Y. Simon, ed., *The Papers of Ulysses S. Grant,* 18 vols. (Carbondale: Southern Illinois University Press, 1967–), 1:243.

2. Helen Nicolay, *Lincoln's Secretary: A Biography of John G. Nicolay* (New York: Longmans, Green, 1949), 195–96.

3. Simon, ed., *Grant Papers,* 10:195.

4. Horace Porter, *Campaigning with Grant* (New York: The Century Co., 1897), 22.

5. John Russell Young quoted in *Chicago Tribune,* Sept. 1, 1885.

6. Sherman to Grant, March 10, 1864, Simon, ed., *Grant Papers,* 10:188.

7. *Personal Memoirs of U. S. Grant* (New York: Charles L. Webster, 1885–86), 2:122; Porter, *Campaigning with Grant,* 26.

8. Lincoln to Grant, April 30, 1864, and Grant to Lincoln, May 1, 1864, Simon, ed., *Grant Papers,* 10:380.

9. Tyler Dennett, ed., *Lincoln and the Civil War in the Diaries and Letters of John Hay* (New York: Dodd, Mead, 1939), 179; Grant to Sherman, April 4, 1864, Simon, ed., *Grant Papers,* 10:253; Grant, *Memoirs,* 2:143.

10. Lincoln to Grant, June 15, 1864, Simon, ed., *Grant Papers,* 11:45; Roy P. Basler et al., eds., *The Collected Works of Abraham Lincoln,* 9 vols. (New Brunswick: Rutgers University Press, 1953–55), 7:393.

11. Grant, *Memoirs,* 2:116.

12. John M. Schofield, *Forty-Six Years in the Army* (New York: The Century Co., 1897), 361–62.

13. John Russell Young, *Around the World with General Grant,* 2 vols. (New York: American News Company, 1879), 2:214, 216–17, 463.

14. Grant to Halleck, Jan. 20, 1863, Simon, ed., *Grant Pa-*

pers, 7:234; James Harrison Wilson, *Under the Old Flag* (New York and London: D. Appleton, 1912), 1:148–49; Adam Badeau to S. L. M. Barlow, Feb. 27, 1863, Henry E. Huntington Library, San Marino, Calif.

15. Howard K. Beale, ed., *Diary of Gideon Welles: Secretary of the Navy Under Lincoln and Johnson,* 3 vols. (New York: W. W. Norton, 1960), 1:387.

16. Lincoln to Grant, July 13, Aug. 9, 1863, Basler, ed., *Collected Works of Lincoln,* 6:326, 374; Grant to Lincoln, Aug. 23, 1863, Simon, ed., *Grant Papers,* 9:195.

17. Grant to David D. Porter, May 25, 1868, ibid., 18:262; Grant to William T. Sherman, June 21, 1868, ibid., 18:292.

18. Grant to Lincoln, July 25, 1864, ibid., 11:309.

19. Grant to Halleck, Aug. 1, 1864, ibid., 11:361.

20. Basler, ed., *Collected Works of Lincoln,* 7:470.

21. Grant to Halleck, Aug. 1, 1864, Simon, ed., *Grant Papers,* 11:358.

22. Lincoln to Grant, Aug. 3, 1864, Basler, ed., *Collected Works of Lincoln,* 7:476.

23. Grant to Halleck, April 22, 1864, Simon, ed., *Grant Papers,* 10:340.

24. Grant to Halleck, April 26, 1864, ibid., 10:356.

25. Halleck to Grant, April 29, 1864, ibid., 10:369.

26. Halleck to Grant, May 2, 1864, ibid., 10:375.

27. Grant to Halleck, July 7, 1864, ibid., 11:185.

28. Charles A. Dana to Grant, July 12, 1864, ibid., 11:230.

29. Grant to Halleck, July 1, 1864, ibid., 11:155.

30. Halleck to Grant, July 3, 1864, ibid., 11:156.

31. Grant to Halleck, June 24, 1864, ibid., 11:124.

32. Memorandum, Aug. 23, 1864, Basler, ed., *Collected Works of Lincoln,* 7:514.

33. Stanton to Grant, Oct. 2, 1864, Simon, ed., *Grant Papers,* 12:303. See Grant, *Memoirs,* 2:375–76; Porter, *Campaigning with Grant,* 314–16.

34. Grant to Washburne, Aug. 16, 1864, Simon, ed., *Grant Papers,* 12:16–17.

35. Grant to Washburne, Sept. 21, 1864, ibid., 12:185.

36. Grant to Lincoln, Jan. 31, 1864, ibid., 13:333–34; Lincoln to Grant, Feb. 1, 1865, Basler, ed., *Collected Works of Lincoln,* 8:252.

37. Grant to Stanton, Feb. 1, 1865, Simon, ed., *Grant Papers,* 13:345.

38. Lincoln to Seward, Feb. 2, 1865, Basler, ed., *Collected Works of Lincoln,* 8:256.

39. John Y. Simon, ed., *The Personal Memoirs of Julia Dent Grant* (New York: G. P. Putnam's Sons, 1975), 141.

40. Simon, ed., *Grant Papers,* 14:64.

41. Stanton to Grant, March 3, 1865, ibid. (cancellations omitted), 14:91.

42. Lincoln to Grant, Jan. 19, 1865, Basler, ed., *Collected Works of Lincoln,* 8:223.

43. Grant to Lincoln, Jan. 21, 1865, Simon, ed., *Grant Papers,* 13:281.

44. Ibid., 13:282.

45. Porter to Sherman, 1866, in *Memoirs of Gen. W. T. Sherman, Written by Himself* 4th ed., 2 vols. (New York: Charles L. Webster, 1891), 2:329.

46. Grant, *Memoirs,* 2:514.

47. Lincoln to Grant, April 7, 1865, Basler, ed., *Collected Works of Lincoln,* 8:392.

48. Simon, ed., *Grant Papers,* 14:484.

For Further Reading

A Bibliography

One: Lincoln and McClellan

STEPHEN W. SEARS

This paper is derived from the author's *George B. McClellan: The Young Napoleon* (1988) and the documentation for that biography, *The Civil War Papers of George B. McClellan: Selected Correspondence, 1860–1865* (1989). Two of the author's campaign studies also stress McClellan's relationship with Lincoln: *To the Gates of Richmond: The Peninsula Campaign* (1992) and *Landscape Turned Red: The Battle of Antietam* (1983).

McClellan's own views of these events will be found in his *Report on the Organization of the Army of the Potomac, and of Its Campaigns in Virginia and Maryland* (1864) and *McClellan's Own Story* (1887). For the nefarious role of McClellan's literary executor in the latter, see Stephen W. Sears, "The Curious Case of McClellan's Memoirs," *Civil War History,* 34 (June 1988): 101–14.

Useful descriptions of McClellan in command by his contemporaries include John G. Barnard, *The Peninsular Campaign and Its Antecedents* (1864); Prince de Joinville, *The Army of the Potomac: Its Organization, Its Commander, and Its Campaign* (1862); Comte de Paris, *History of the Civil War in America,* 4 vols. (1875–88); James B. Fry, "McClellan and His 'Mission'," *The Century,* 48 (October 1894): 931–46; William F. Biddle, "Recollections of McClellan," *United Service Magazine,* 11 (May 1894): 460–69; and A. K. McClure, "Lincoln and McClellan," in *Abraham Lincoln and Men of War-Times* (1892).

Among the useful modern studies of aspects of McClellan's military role are two by T. Harry Williams, *Lincoln and His Generals* (1952) and *McClellan, Sherman, and Grant* (1962); Herman Hattaway and Archer Jones, *How the North Won: A Military History of the Civil War* (1983); Archer Jones, *Civil War Command & Strategy: The Process of Victory and Defeat* (1992); Michael C. C. Adams, *Our Masters the Rebels: A Speculation on Union Military Failure in the East, 1861–1865* (1978); Joseph L. Harsh, "On the McClellan-Go-Round," *Civil War History,* 19 (June 1973): 101–18; and two institutional studies by Edward Hagerman, "The Professionalization of George B. McClellan and Early Civil War Field Command: An Institutional Perspective," *Civil War History* 21 (June 1975): 113–35, and *The American Civil War and the Origins of Modern Warfare: Ideas, Organization, and Field Command* (1988).

The key role of military intelligence in McClellan's thinking is examined in Edwin C. Fishel, "Pinkerton and McClellan: Who Deceived Whom?," *Civil War History,* 34 (June 1988): 115–42.

Every Lincoln biographer perforce deals with the president's relationship with General McClellan. Probably the most anti-McClellan is John G. Nicolay and John Hay, *Abraham Lincoln: A History,* 10 vols. (1886–90). The most pro-McClellan is James G.

Randall, *Lincoln the President: Bull Run to Gettysburg* (1956). Other contributors to this subject are: Colin R. Ballard, *The Military Genius of Abraham Lincoln* (1926); Maurice Frederick, *Statesmen and Soldiers of the Civil War: A Study of the Conduct of War* (1926); and Herman Hattaway and Archer Jones, "Lincoln as a Military Strategist," *Civil War History,* 26 (1980): 293–303. The president's war views are perceptively examined in James M. McPherson, "Lincoln and the Strategy of Unconditional Surrender," in Gabor S. Boritt, ed., *Lincoln, the War President: The Gettysburg Lectures* (1992). The core of the president's military thought is, of course, to be found in Roy P. Basler et al., eds., *The Collected Works of Abraham Lincoln,* 9 vols. (1953–55).

Two: Wilderness and the Cult of Manliness: Hooker, Lincoln, and Defeat

MARK E. NEELY, JR.

"There is no comfortable way of reading military history," warned John Bigelow, Jr., the famous historian of the battle of Chancellorsville. "Whoever expects to follow a campaign reclining in an easy chair with a book in one hand and a cigar in the other is doomed to disappointment." The statement is true even for a campaign as often studied and written about as Chancellorsville. For this brief article, no equal to Bigelow's massive and much-admired work, I traveled to Virginia and spent two days touring the fields. And I have drawn on the resources of several university research libraries merely to gather the salient published works on Chancellorsville. But I have been forced to leave most manuscript sources for another day.

The literature on the campaign is extensive and of high quality, and it touches a surprisingly wide range of familiar themes in American history. The first great decade for Civil War studies, the

1880s, was also the first great decade of Chancellorsville studies. Theodore A. Dodge's book *The Campaign of Chancellorsville* (1881) mustered a relentless assault on Hooker, whom Dodge depicted (p. 228) as a cowardly braggart like Falstaff. Hooker's defenders weighed in immediately afterward. Samuel Penniman Bates, in a book little marked by acrimony, gave Hooker's side of the story in *The Battle of Chancellorsville* (1882). Bates places responsibility for the defeat on O. O. Howard, taking care not to blame the men of the XI Corps. He based his revised estimate of the battle on the emerging availability of Confederate sources and estimated the positions and movements of the enemy as well as the Federal forces. Abner Doubleday's volume on the battle in the Campaigns of the Civil War series appeared the next year—*Chancellorsville and Gettysburg* (1882)—and offered a more balanced assessment. He blamed Hooker for the initial pullback from the edge of the Wilderness; O. O. Howard for not preparing for a flank attack; and Hooker's injury for the irresolute failure to send in the tens of thousands of unengaged reinforcements still available to the Union army on May 3.

A relevant work in the same Scribner's series was Andrew A. Humphreys, *The Virginia Campaign of 1864 and 1865,* published in 1885. Here a reader could find the Wilderness invoked as an excuse for Ulysses S. Grant's defeat in the battle of the Wilderness, fought almost exactly a year after the battle of Chancellorsville. Every third page of the description of Grant's Wilderness campaign builds the excuse into a towering apology of leaves and thickets: "Passing through the same kind of entangled wood found everywhere"; "his front line being so entangled in the wood as not to admit of ready handling"; the "same close underbrush . . . aggravated in the swampy parts"; "how bewildering the dense forest growth was"; "through the forest, undergrowth, and swamps"; "concealed by the dense wood"; and the "advance in this direction was through woods with matted undergrowth, and the progress was very slow" (pp. 27–46). In the end Humphreys concluded:

> I have gone into more detail in the account of this battle than I shall undertake to give of those that are to follow, chiefly because it may serve to show what difficulties were encountered by the forces engaged in it, owing to the character of the field on which it took place. Some of its features were found in other of the battle-grounds of the two armies; but, so far as I know, no great battle ever took place before on such ground [p. 55].

Realizing the statement might sound preposterous to those who had fought near that same field in 1863, as Humphreys himself had, the author added a footnote: "The same ground occupied by the Army of the Potomac in the vicinity of Chancellorsville in the Spring of 1863 was either open or in woods chiefly of ordinary character with but little undergrowth" (p. 55n). Yet a check of the battle reports submitted to Humphreys after Chancellorsville by his own subordinates reveals one, for example, from Colonel Jacob G. Frick of the 129th Pennsylvania, which explained that the movement of his troops in an attack at Chancellorsville was "in an undergrowth that, from its density, made the movement peculiarly difficult" (*OR,* I, 25, pt. 1. p. 554)!

Cavalry general Alfred Pleasonton, writing for the famous *Battles and Leaders of the Civil War* series (edited by Robert Underwood Johnson and Clarence Clough Buell, 4 vols., 1888), argued that Hooker's loss of initiative caused the defeat *because* it left the Union army in the Wilderness. "A march of three or four miles would take us out of the woods into a more open country, where we could form our line of battle, and where our artillery could be used to advantage" (3:174). When Hooker ceased to maneuver and acted as though he wanted to fight a defensive battle against Lee, the Union army was, according to Pleasonton, who had scouted the region, left in this situation.

> To the east, toward Fredericksburg, the woods were thick for three or four miles; to the south, toward Spotsylvania Court House, the woods extended about the same distance; to the west, from Hazel Grove, the same condition of things existed; while the country between Chancellorsville and the Rappahannock River, in our rear, was rough, broken, and not at all suitable for the operations required of an army. The position of the army at Chancellorsville extended about three miles from east to west in the narrow clearings, which did not afford sufficient ground to manœuvre an army the size of the Army of the Potomac. Besides this, we were ignorant of what might be going on outside of this cordon of woods, and were giving the enemy every opportunity to take us at a disadvantage.[3:175]

Perhaps no Civil War battle was more bitterly argued about in literature generated by the northern side. That was due to a feeling now happily departed and difficult to comprehend: intense anti-immigrant sentiments fueled the fires of controversy when Hooker's defeat was blamed on the "German corps," the Eleventh, that crumbled under Stonewall Jackson's flanking attack. To recover a sense of the atmosphere surrounding that controversy, one needs to remember that the Civil War came some five years after the formation of the most powerful anti-immigrant party in this nation's history, for a time a major competitor of the Republican party as the fastest growing party in the country, the Know-Nothing party. Much of the postwar literature about the battles and leaders was written in a continuing climate of anxiety among some Americans about the waves of immigration to these shores. Thus Francis Amasa Walker's *History of the Second Army Corps in the Army of the Potomac* (1886) heaped scorn on the pell-mell flight of the German-Americans on Hooker's right. Admitting that native-born Americans also sometimes were put to flight, Walker nevertheless said:

> I never saw an American so frightened as to lose his senses, though I have seen thousands of the natives of Columbia leave one battle-field or another in the most dastardly manner. But if an American is mean enough to abandon the line, it is always done cooly and collectedly. Indeed, he will exhibit a degree of engineering skill in getting out of a fight under cover which would do credit to a member of the topographical staff. [p. 229n]

The landmark Chancellorsville study that grew out of that controversy was Augustus Choate Hamlin, *The Battle of Chancellorsville: The Attack of Stonewall Jackson and His Army upon the Right Flank of the Army of the Potomac at Chancellorsville, Virginia, on Saturday Afternoon, May 2, 1863* (1896). Hamlin set out to vindicate the rank and file of the Eleventh Corps, and in doing so laid blame for the defeat on its leaders, especially West Point-trained generals. He singled out Hooker and Warren, among others. He said little about Howard, also a West Pointer, but certainly left the impression that he ignored warnings and left his corps prey to Jackson's attack. This little volume, based on five years' research and three trips to the battlefield in successive years, also represented the culmination of the wilderness argument for Union defeat. Hamlin, seeking causes for the collapse of the Eleventh Corps other than indiscipline, cleverly fit the Union defeat and the arrogance and pride of the West Pointers into the oldest American military myth, the myth of Braddock's defeat. The richly descriptive passages with their images of butternut-clad Confederates in thick forests have already been quoted in the text of this article. Faint echoes of Hamlin's work have appeared in the literature ever since, right down to the most recent work, published in 1992, in which the author, Ernest B. Furgurson, describes the Confederates as "more woods-wise, and unafraid of ripping their clothes and themselves crashing through the brush" (p. 337).

James Beale offered an original interpretation in a little paper read before the United Service Club of Philadelphia in 1888, *Chancellorsville* (1892). By seizing on all reports of reluctance to go forward and rumors of eagerness to flee, he hinted that the demoralization induced in the Army of the Potomac by the disaster of Fredericksburg had not been as fully reversed as Hooker and many commentators thought. Such an argument flew in the face of literally hundreds of colonels' after-battle reports available in the *Official Records,* with their formulaic assertions of dutiful effort and willingness to try again, and it had to contend with ordinary patriotic feeling as well. In the end the argument did not prove influential. Despite the advent, in current times, of studies of battles from the standpoint of the common soldier rather than of headquarters, the theory has not yet really been tested. It probably is not true, but it deserves more of an examination than it has received to date. Another factor affecting Union soldiers' morale and stamina at Chancellorsville, a controversial order, widely disobeyed and protested, to carry eight days' rations for the campaign, has likewise gone without systematic examination from the point of view of the common soldier who had to carry the extra rations. Written entirely from the headquarters' point of view, Edward Hagerman's *The American Civil War and the Origins of Modern Warfare: Ideas, Organization, and Field Command* (1988) pronounces the experiment "a logistical success" (pp. 72–73).

Of lesser note is A. H. Nelson's semi-literate, confused, and partly plagiarized little book, *The Battles of Chancellorsville and Gettysburg* (1899). Captain Nelson, who was present at the battles in the 57th Pennsylvania, stoutly defended Hooker by blaming defeat on Howard, who "ought to have been tried by drumhead court-martial and shot before the army recrossed the river for wilful disobedience of orders" (p. 76). It is true that other generals suffered more for less serious failures than Howard. Nelson hints at various conspiracies and accuses Halleck of scheming to undo Hooker. The book is a monument only to the ability of discussions of Chancellorsville to arouse passions (and proof that ignorant Pennsylvania timbermen ought not to write books).

John Bigelow, Jr.'s *The Campaign of Chancellorsville: A Strategic and Tactical Study* (1910) is a work of near-legendary status among Civil War battle studies. Even this balanced history gives the anti-immigrant interpretation some credit, noting that "our German population comes to our country to escape from military service, and for this reason is not representative of the military population of Germany" (p. 480). But the prestige of German arms in Bigelow's era—he noted the distinguished military tradition of Frederick the Great, Blücher, and Moltke—and Bigelow's good sense were too great to lead him to give it much credence. Instead, he opted for an interpretation that faulted Hooker for having in mind from the start a defensive battle fought, in a phrase that became a steady refrain in the general's despatches, "on his own ground." He also faulted Howard and General Slocum for the faults that Hamlin had found in them: they thought "that the forest about them was inpenetrable to troops except on the roads, and . . . that they were abundantly able to hold their position against any force which the nature of the ground in their front would enable the enemy to bring against them, and that to fall back would have some of the demoralizing effect of a retreat" (p. 258). Thus they did not alter the alignment of the right flank or strengthen it greatly. Bigelow probably put more emphasis on Hooker's health than any other writer to date—more even than Hooker himself, who remained true to his stoic masculine ideal.

Bigelow gave some credence to the wilderness theory, though terrain was not the principal thrust of his argument despite the resonance of "on our own ground" throughout the book. He even picked up from Hamlin the theme that the "forests did not prevent the Confederates as they did the Federals from deploying off the roads and marching in line. The Confederates moved through the Wilderness in every direction and in every kind of formation. They were better woodsmen than the Federals, and better acquainted with the terrain, or better supplied with guides" (p. 476). Mainly he faulted Hooker's "irresolution" (p. 477) and his lack of "the imagination necessary to keep before his mind the changing positions of troops out of his sight. His mental vision was practically limited by his physical vision, and he had apparently no training or faculty for making war on a map" (p. 482). Finally, Bigelow stated that "No greater mistake was made during the campaign than Hooker's final one of recrossing the Rappahannock. Lee was about to play into his hands by attacking him on his own ground; the condition on which his plan of operation was based was at last to be realized, when he weakly retired from the contest" (p. 482).

Colonel J. E. Gough's *Fredericksburg and Chancellorsville: A Study of the Federal Operations* (1913), written by a British soldier for British soldiers, has a certain period charm about it and an eye for the practical. He condemns the eight-day-ration scheme, for example, and he reads Hooker's orders very closely to prove their tone of indecisiveness. In the end he said that Hooker's loss of initiative lost the battle, and—a portent of things to come—he ended the book with a quotation from Lincoln which he pronounced "sound" strategic thinking.

Strangely, the advent of the Great War in Europe did not alter interpretations of Chancellorsville except perhaps in making it seem irrelevant and an unpromising object of military study. In 1915 David Gregg McIntosh, who had been a Confederate colonel of artillery in Jackson's flanking movement, wrote a balanced brief account, *The Campaign of Chancellorsville* (Richmond, Va.: William Ellis Jones' Sons, n.d.), which emphasized the virtues of attack and the vices of losing initiative. It would have been "criminal" for a general who knew "the thinness of Lee's lines" to fail to attack after May 3 (p. 56); Hooker must have been in a stupor induced by his wound; and Stonewall Jackson "was a thorough believer in the military principle which reckons the advantage on the side of the attacking party" (p. 3).

A long hiatus intervened in the scholarship of the Chancellorsville campaign. Bigelow's detailed study was daunting and seemingly definitive. Chancellorsville's fame for demonstrating the virtues of maneuver on Lee and Jackson's part may have made it seem almost quaint after the static trench warfare on the Western Front. And eventually new theories of the reasons for Union victory in the Civil War emerged in which Chancellorsville did not fit—not because it was a Union defeat or involved daring maneuver but because the battle did not seem to involve the salient new topics: "total war," Jomini and Clausewitz, and the harbingers of twentieth-century industrial wars of indiscriminate destruction, "attrition," and "unconditional surrender." For forty-five years, in fact, no specialized works on the battle appeared, but in 1958 World War I veteran Edward J. Stackpole's, *Chancellorsville: Lee's Greatest Battle,* helped to revive the debate. A somewhat amateurish work marked by occasionally banal writing (Stackpole characterized Lee, for example, as "a very human person as well as a proven genius in war" (p. 360)), the book maintained the traditional emphasis on Hooker's loss of "momentum" (p. 362) and the necessity "to get clear of the dense wilderness into open and maneuverable country." He accused Hooker in the bitterest language of "almost criminal military negligence" in failing "to secure his right flank" (p. 364), for example. He also attributed to Hooker a "fear psychosis of some sort, although that is something for the psychiatrists to explain" (p. 365). And thus Stackpole brought the traditional emphasis on Hooker's failure to maintain the initiative into the modern language of psychology.

The most recent work is Ernest B. Furgurson's *Chancellorsville 1863: The Souls of the Brave* (1992). Characteristic of our age of military historical writing, the principal new sources cited are diaries and letters of common soldiers, many from manuscript collections, but the descriptions and judgments of command decisions follow the previous literature. The views of the common soldiers ultimately add nothing except lively anecdotes, as they are not used systematically as a check on the viewpoint of the officers, on the woodsmanship of the Union army, or of the effects of the eight days' ration order. Furgurson's ultimate assessment relies on Bigelow and other able predecessors to fault Hooker for losing initiative and to praise Lee for recognizing Hooker's loss of confidence and for taking many chances that could not have been taken in the face of an unflinching opponent.

The culmination of the psychological interpretation of Chancellorsville, which places most emphasis on the mysterious loss of initiative in Hooker, is the treatment of the battle that appears in Michael C. C. Adams, *Fighting for Defeat: Union Military Failure in the East, 1861–1865* (1992; orig. pub. as *Our Masters the Rebels: A Speculation on Union Military Failure in the East, 1861–1865* in 1978). Adams's influential argument in the book suggests that the Army of the Potomac, because of the accident of losing its first battle, Bull Run, to the Confederates, and because of nineteenth-century ideas that associated prowess in war with pastoral lifestyles more prevalent in myths associated with the South, had an inferiority complex. It beautifully explains Hooker's case as traditionally interpreted. As Adams put it, Hooker "in the position of responsibility, fell a victim to the ghost of past inferiority feelings" (p. 143). Sadly, despite his acute awareness of the belief of Civil War soldiers in pre-industrial sources of military ability,

Adams did not offer further "speculation" on the role of terrain in the campaign and in the historiography of the campaign.

My invitation to write this essay on Hooker, Lincoln, and defeat provided me my first opportunity to study the historiography of a Civil War battle. I have been impressed with the revealing nature of this historiographical enterprise and with the richly varied literature available. There is scant historiographical work on the war, and opportunity beckons. In the case of Chancellorsville itself, surely the psychological/momentum interpretation has by now given Civil War students all it has to offer, and the battle should be examined from other perspectives. I find it remarkable that authors for over a hundred years could cite Pleasonton's argument about getting out of the Wilderness without ever pausing to consider what that means to the powerful myth that adaptiveness to hostile terrain supplied the peculiar essence of American military abilities. Likewise, the striking examples of language in the battle reports and in the heroic behavior of officers and men in the field, suggestive of the need to explore the applicability of the code of manliness to American military ideals, show one way in which that great struggle in Virginia's forests was a laboratory of human behavior under extreme conditions of stress. There remains plenty of opportunity for research in that verdant lab.

Three: "Unfinished Work": Lincoln, Meade, and Gettysburg

GABOR S. BORITT

Gettysburg is the most written about battle of American history. Richard Allen Sauers, *The Gettysburg Campaign, June 3–August 1, 1863* (1982) remains the best bibliographical source, and includes the concluding part of the campaign. The bibliography needs updating.

Edwin B. Coddington continues as the finest expert on the subject, though his majestic work, too, is now dated, both by its perspective, identified in his book's subtitle, and by the outpouring of newer research. *The Gettysburg Campaign: A Study in Command* (1968) devotes its final detailed chapter to "Retreat and Pursuit." Unlike many earlier works, it emphasized the difficulties Meade faced and became his most capable champion. Coddington presaged the views of his magnum opus in "The Strange Reputation of General Meade: A Lesson in Historiography," *Historian,* 23 (Feb. 1961): 145–66, and "Lincoln's Role in the Gettysburg Campaign," *Pennsylvania History* 34 (1967): 250–65. These articles showed the president playing a crucial part in what Coddington saw as the unfairly anti-Meade "verdict of history," though this historian ultimately faulted "an American trait which thinks of victory as nothing less than decisive and surrender as nothing but unconditional." (The quotes are from *Pennsylvania History,* 265, and the *Historian,* 165.) Coddington's influence is well illustrated by Amy J. Kinsell's fine 1992 Cornell University dissertation, " 'From these honored dead': Gettysburg in American Culture, 1863–1938," which won the Allan Nevins award as the best American doctoral dissertation in history in 1992. It talks of Lee's "escape" with the word in quotation marks to suggest irony. See also, for example, John W. Schildt, *Roads from Gettysburg* (1979), a gold mine of information that needs careful handling. Students await eagerly Kent Masterson Brown's forthcoming book on Lee's retreat.

Harry W. Pfanz, "The Gettysburg Campaign after Pickett's Charge," *Gettysburg* 1 (July 1989): 118–24, is a fine brief overview.

Among the pre-Coddington books, two are still readily available to readers. Edward J. Stackpole's *They Fought at Gettysburg* (1956) considered the climax of the battle's aftermath in a breathy chapter entitled, "A Lost Opportunity to End the War." However, Glenn Tucker's *High Tide at Gettysburg* (1958) concluded that the question of whether an attack by Meade might have succeeded will, of course, "never be answered" (p. 387). Then, two sentences later, he answered thus: "It seems unlikely Meade could have done much better than Burnside did at Fredericksburg"—one of the great Union defeats of the Civil War. The dispute goes all the way back to July of 1863, and carries whiffs of being either on the side of the military men, a noble thing to do, or the politicians, something always more questionable in popular culture. Frank A. Haskell fired the first shot against "the formidable warriors from the brothels of politics" who passed judgment "at safe distances from guns, in the enjoyment of a lucrative office, or of a fraudulently obtained government contract, surrounded by the luxuries of their own firesides, where mud and flooding storms, and utter weariness never penetrate . . . ," in *The Battle of Gettysburg* (1910), p. 196. Though Haskell wrote in the summer of 1863, and in time historians unearthed the revered Lincoln's central role on the side of politicians, the dichotomy noted above never quite disappeared.

Mary Chesnut smiled bitterly at this general problem while commenting on the capture of Confederate president Jefferson Davis in 1865: " 'Knights-errant themselves, when they conquered giants, slew the giants. Quarter was only for other knights-errant who knew the courtesy and the laws of battle'. . . . Which, being interpreted, means Jeff Davis must pay" and not Lee, or the other Confederate knights-errant (C. Vann Woodward, ed., *Mary Chesnut's Civil War* (1981): 823). In the context of the retreat and pursuit after Gettysburg, to question Meade is to defend Lincoln and his legend—even if his confusions and exaggerations are noted. Furthermore, to question Meade's competence is to question Lee's also, and his legend, which outstrips even Lincoln's among students of military history. For if Meade performed inadequately, Lee's achievement becomes the less remarkable.

But to return to the inimitable Haskell, his work also provides a fine example of the glorious and mountainous rubble of contemporary Gettysburg letters, diaries, newspaper and magazine accounts, memoirs, and more, published and unpublished, that cannot be listed here. Of course, much remains undiscovered in private holdings and, no doubt, in public ones.

The Official Records are indispensable, though like all official records they contain a solid element of self-serving materials. The pertinent volumes for Gettysburg are Series 1, volume 23, parts 1, 2, and 3. The congressional investigations of 1864, Meade's "second battle of Gettysburg," can also be followed in public documents: *Report of the Joint Committee on the Conduct of the War,* 38 Cong., 2 sess. (Washington, 1865).

From the perspective of this essay, it is painful how much more information is available about Lincoln than Meade, how comparatively little we know about the latter. The general needs a new biography. A fine brief sketch is Noah André Trudeau's " 'I Have a Great Contempt for History,' " *Civil War Times Illustrated,* 30 (Sept.–Oct. 1991): 31–40. It is worth the effort to look up Frederick Bernays Wiener, "Decline of a Leader: The Case of General Meade," *Infantry Journal,* 45 (Nov.–Dec. 1938, and Jan.–Feb. 1939). Freeman Cleaves, *Meade of Gettysburg* (1960), is a good biography. The best printed source material is George Meade, *The Life and Letters of George Gordon Meade,* George Gordon Meade [grandson of the general], ed., 2 vols. (1913). However, the careful student needs to consult the original manuscripts, since the editing leaves very much to be desired by modern scholarly standards. The same is true of numerous other older collections.

For Lincoln the best source is the Roy P. Basler edition of *The Collected Works of Lincoln,* 11 volumes total, including the index (1953–80), noted in the bibliographies of the other chapters in this book. Among other volumes so listed, Kenneth P. Williams, *Lincoln Finds a General,* vol. 2 (1949), is relentlessly hostile to Meade, even labelling him, repeatedly, "untruthful," pp. 732, 733, 738. T. Harry Williams, *Lincoln and His Generals* (1952) is also strongly critical of Meade. Herman Hathaway and Archer Jones, "Lincoln as a Military Strategist," *Civil War History,* 26 (1980): 293–303, deserves careful consideration, as does James M. McPherson, "Lincoln and the Strategy of the Unconditional Surrender," in Gabor S. Boritt, ed., *Lincoln, the War President: The Gettysburg Lectures* (1992): 29–62.

Shelby Foote, *The Civil War: A Narrative,* vol. 2: *Fredericksburg to Meridian* (1962) contains a long and sprightly retelling of Lee's retreat and Meade's pursuit. Garry Wills, *Lincoln at Gettysburg: The Words That Remade America* (1992) is a provocative work on this subject. Stephen B. Oates, *With Malice Towards None: The Life of Abraham Lincoln* (1977) continues as the standard popular biography, even as David Herbert Donald's forthcoming book is awaited with much anticipation. The best overview is Mark E. Neely, Jr., *The Last Best Hope of Earth: Abraham Lincoln and the Promise of America* (1993).

Four: Lincoln and Sherman

MICHAEL FELLMAN

Despite the abundance of secondary sources about Sherman and about the war in the West, any serious student of the man should look to the archives. Luckily the two major collections are both available in microfilm, the Sherman Papers at the Library of Congress, and the Sherman Family Papers at Notre Dame University Archives. The former collection is strong in Sherman's public career; although it contains many unimportant incoming letters, it does include the life-long correspondence Sherman had with John Sherman. In other collections at the Library of Congress relationships with the Ewing family and with many army officers are detailed. In the Notre Dame collection, spelled out in great detail, are to be found Sherman's relations with his family, particularly his life-long contest with his wife, Ellen Ewing Sherman. As for printed primary sources, for the war years, the *Official Records of the Rebellion* are of course of primary importance, as are Roy Basler's *Collected Works of Abraham Lincoln* (1953), *Supplements* (1974 and 1980). John Y. Simon's superbly edited *Papers of U. S. Grant* detail Sherman's long association with Grant, not only for the war years, but for the postwar years when both men led the fight against the Plains Indians. Later volumes will no doubt demonstrate the many strains which developed in this relationship after Grant went off to his unhappy tenure at the White House. Two volumes of Sherman correspondence, one with his brother, the other with his wife, must be read with attention to poor and misleading editing. They are Mark A. DeWolfe Howe, ed., *Home Letters of General Sherman* (1909) and Rachel Sherman Thorndike, ed., *The Sherman Letters: Correspondence Between General and Senator Sherman from 1837 to 1891* (1894).

For a full and up-to-date listing of the hundreds of other archival collections of Sherman materials, as well as for the secondary literature, including Sherman's many postwar effusions in the popular magazines of his day, the reader is referred to the excellent bibliography in John F. Marszalek's thoroughly researched and presented biography, *Sherman: A Soldier's Passion for Order* (1993), pp. 587–611. In addition to Marszalek's modern study, Lloyd Lewis, *Sherman, Fighting Prophet* (1932), is one of the classics of American biography. Somewhat dated, particularly in its discussions of black troops and of Reconstruction, which amounts to a racist defense of Sherman's racist policies, the Lewis biography is also quite penetrating as a character study, and rather sophisticated about the political culture in which Sherman was operating. James M. Merrill, *William Tecumseh Sherman* (1971) is interesting particularly for Sherman's family life, though it is annoyingly unannotated. Basil H. Liddell-Hart, *Sherman, Soldier,*

Realist, American (1929), is in some respects interesting about military matters, though it reveals more about Liddell-Hart than about Sherman. Charles Royster's *The Destructive War, William Tecumseh Sherman, Stonewall Jackson and the Americans* (1991), winner of the Lincoln Prize, is a strikingly narrated discussion of two of the war's leading figures. Sherman's own two-volume *Memoirs,* originally published in 1875 and frequently reprinted, is beautifully written and artfully argued special pleadings. It should be used with extreme care as history, but at the same time it amounts to a complex and argumentative historical document if read for what it is.

Many other monographs deal with important facets of Sherman's life and career, for the war years and the years after the war, and for his family life. For the Atlanta campaign of the spring and summer of 1864, see Albert Castel's Lincoln Prize-winning, *Decision in the West* (1992). For earlier military events see Margie Riddle Bearss, *Sherman's Forgotten Campaign: The Meridian Expedition* (1987), and James Lee McDonough, *Shiloh—In Hell Before Night* (1977). For campaigns after the fall of Atlanta, see Joseph T. Glatthaar's imaginative social history, *The March to the Sea and Beyond: Sherman's Troops in the Savannah and Carolina Campaigns* (1985); John G. Barrett, *Sherman's March Through the Carolinas* (1956); and Marion B. Lucas, *Sherman and the Burning of Columbia* (1976). On one of Sherman's great wartime obsessions see Marszalek, *Sherman's Other War: The General and the Civil War Press* (1981).

Two studies of Sherman's family are of special importance. Joseph T. Durkin's *General Sherman's Son* (1959), deals with Thomas Sherman, who became a Jesuit priest in 1878, much to his father's dismay. Anna McAllister, *Ellen Ewing: Wife of General Sherman* (1936), helps underline the pivotal role Ellen Ewing Sherman played in her husband's public as well as private life.

Several studies reconstruct aspects of Sherman's postwar life and career. Notable among them are Richard A. Andrews, "Years of Frustration: William T. Sherman, the Army, and Reform, 1869–1883," Ph.D. dissertation, Northwestern University, 1968; Robert G. Athearn, *William Tecumseh Sherman and the Settlement of the West* (1956); Robert M. Utley, *Frontier Regulars: The United States Army and the Indians, 1866–1891* (1973); and, on the Whittaker case, a sorry racist episode late in Sherman's career, Marszalek, *Court Martial: A Black Man in White America* (1972).

For those who might desire even more discussion of Sherman in the future, I will weigh in with a full-length treatment of the personality and cultural fit of this legendary figure, in a book entitled *Citizen Sherman*. Whatever I may add, however, there will never be a definitive study of this protean and troubling man, a man who truly bestrode the Colossus.

Five: Grant, Lincoln, and Unconditional Surrender

JOHN Y. SIMON

Because Abraham Lincoln and Ulysses S. Grant met rarely but corresponded frequently, their relationship can be readily approached through their mail and telegraph communications printed in Roy P. Basler et al., eds., *The Collected Works of Abraham Lincoln,* 9 vols. (1953–55) and John Y. Simon, ed., *The Papers of Ulysses S. Grant,* 18 vols. to date (1967–). In his impoverished final days, while dying of cancer, Grant wrote a classic military autobiography: *Personal Memoirs of U. S. Grant,* 2 vols. (1885–86). His wife later prepared *The Personal Memoirs of Julia Dent Grant* (1975). Staff officer Horace Porter, *Campaigning with Grant* (1897; reprinted 1961), provides a meticulously detailed, accurate, and readable narrative of the last year of the war from a headquarters perspective.

British general Colin R. Ballard, *The Military Genius of Abraham Lincoln* (1926; reprinted 1952), opened a debate continued by T. Harry Williams, *Lincoln and His Generals* (1952). Williams argued that Lincoln monitored Grant's military operations in the final year of the war and retained control of grand strategy. Kenneth P. Williams, *Lincoln Finds a General: A Military Study of the Civil War,* 5 vols. (1949–59), placed more emphasis on Grant's ability. The author's death ended his series, however, before he could extend his analysis beyond the campaigns of 1863.

Another British officer, Major General J. F. C. Fuller, *Grant & Lee: A Study in Personality and Generalship* (1933; reprinted 1957), began a modern revival of Grant's military reputation, effectively advanced by Bruce Catton, *Grant Moves South* (1960) and *Grant Takes Command* (1969). Catton, *A Stillness at Appomattox* (1953), remains a classic account of the 1864–65 campaign of the Army of the Potomac, and his *U. S. Grant and the American Military Tradition* (1954) captures the essential man in fewer than 200 pages of large print.

Recent Grant biographies include William S. McFeely, *Grant: A Biography* (1981) and Brooks D. Simpson, *Ulysses S. Grant and the Politics of War and Reconstruction, 1861–1868* (1991). These provocative studies are complementary: McFeely's Grant can do nothing right, Simpson's can do nothing wrong.

Charles B. Strozier, *Unconditional Surrender and the Rhetoric of Total War: From Truman to Lincoln* (1987), relates Grant's concept of unconditional surrender to the decision to drop the atomic bomb on Japan but concedes that unconditional surrender fit the circumstances of the Civil War. James M. McPherson, "Lincoln and the Strategy of Unconditional Surrender," defends Lincoln's refusal to negotiate with Confederates in Gabor S. Boritt, ed., *Lincoln the War President: The Gettysburg Lectures* (1992).

Contributors

Gabor S. Boritt, Director of the Civil War Institute and Fluhrer Professor at Gettysburg College, is author of *Lincoln and the Economics of the American Dream* (1978; University of Illinois Press paperback, 1994). His most recent book is *Lincoln, the War President: The Gettysburg Lectures* (1992; Oxford paperback, 1994). He is currently working on a study of the Battle of Gettysburg.

Michael Fellman is Professor of History at Simon Fraser University in British Columbia. His most recent book is *Inside War: The Guerrilla Conflict in Missouri During the American Civil War* (1989, Oxford paperback, 1990). He is currently completing a book entitled *Citizen Sherman,* to be published by Random House in 1995.

Mark E. Neely, Jr., John Francis Bannon Professor of History and American Studies at Saint Louis University, won the Pulitzer Prize for *The Fate of Liberty: Abraham Lincoln and Civil Liberties* (1991, Oxford paperback, 1992). His most recent book is *The Last Best Hope of Earth: Abraham Lincoln and the Promise of America* (1993). He is writing a book on the Civil War and the two party system.

Stephen W. Sears is the leading expert on George B. McClellan and the author of three books focused on him, including the full-scale biography: *George B. McClellan: The Young Napoleon* (1988) and *To the Gates of Richmond: The Peninsula Campaign* (1992). He is currently making a study of the battle of Chancellorsville.

John Y. Simon, the leading expert on Ulysses S. Grant, Executive Director of the Grant Association, and Professor of History at the Southern Illinois University, Carbondale, is the editor of eighteen volumes of *The Papers of Ulysses S. Grant* (Southern Illinois University Press, 1967–), as well as of *The Personal Memoirs of Julia Dent Grant* (1975; Southern Illinois University paperback, 1988). While completing work on the Grant Papers, Simon continues to write about Abraham Lincoln.